4—

# THE GREAT PIKES PEAK GOLD RUSH

4-18-86

We hope you will return
to the Pikes Peak area.

Ellen Cesarne
Dic Johnson

# *The Great Pikes Peak Gold Rush*

*By*

ROBERT L. BROWN

THE CAXTON PRINTERS, LTD.
CALDWELL, IDAHO
1985

Library of Congress Cataloging in Publication Data

Brown, Robert Leaman, 1921–
    The great Pikes Peak gold rush.

    Includes index.
        1. Pikes Peak Region (Colo.) — Gold discoveries.
    2. Gold mines and mining — Colorado — Pikes Peak Region —
History — 19th century. 3. Pikes Peak Region (Colo.) —
History. I. Title.
    F782.P63B76    1985        978.8'56        85-5767
    ISBN 0-87004-311-0
    ISBN 0-87004-323-4 (pbk.)

Printed and bound in the United States of America by
The Caxton Printers, Ltd.
Caldwell, Idaho 83605
143101

For
Becky Zombola, who was born
in the shadow of Pikes Peak

# TABLE OF CONTENTS

# LIST OF ILLUSTRATIONS

♋

# ACKNOWLEDGMENTS

DURING THE PREPARATION of this book, the author has incurred many debts. First of all, particular gratitude is due my wife, Evelyn McCall Brown, who was my companion on nearly all of the field trips, my source of advice on sentence structure, photographic angles, and for help with proofreading. Louisa Ward Arps and Freda and Francis Rizzari gave unstintingly of their time, skills, and knowledge in proofreading the final manuscripts. Their competence and many suggestions have been invaluable.

Most of the early photographs came from the fine collections of Freda and Francis Rizzari, Frances and Dick Ronzio, and from the late Jo and Fred Mazzulla. Others are from the Nancy and Ed Bathke collection, from Sid Squibb, and the Gilpin County Historical Society. The photographic collections at the Western History Department of the Denver Public Library and the library of the Colorado State Historical Society supplied many other fine pictures unobtainable elsewhere. George Foott of Littleton, able artist of many Colorado scenes, researched and executed the excellent likeness of the Wind Wagon.

Most of the actual research was done at the library of the Colorado State Historical Society and at the Western History Department of the Denver Public Library, where Mrs. Eleanor Gehres and her staff were of particular assistance. The books that I consulted at these institutions are included within the bibliography. Additionally, both institutions made available their microfilms of the *Central City Register Call*, the *Rocky Mountain News*, the *Fairplay Flume*, the *Colorado Miner*, the *Denver Tribune*, the *Denver Republican*, and the *Denver Times*.

During research in the field, Fletcher Birney of Castle Rock located and hiked with me on the Smoky Hill Trail. Earl and Barbara Boland were amiable companions on the hikes along the Big Hill Wagon Road and on the Apex and Gregory Toll Road. Dr. Henry Thode of Fort Collins led us along many miles of the Overland Trail. Ed and Nancy Bathke hiked with us twice along the original Ute Trail. Dick Akeroyd of the Denver Public Library made numerous trips with me to the old site of Montgomery. John Adams of Abraham Lincoln High School made several trips with me to the South Park region. Marian and Dick Ramsey showed us many historic sites in the Breckenridge area.

To all of these good friends and institutions, my heartfelt gratitude.

Robert L. Brown
Denver, Colorado, 1985

# THE GREAT PIKES PEAK GOLD RUSH

# 1.

## THE BARRIER LAND

COLORADO, AMERICA'S CENTENNIAL STATE, is a land of both incredible beauty and starkness. Its eastern geography includes vast rolling stretches of the Great Plains, inclining imperceptibly upward from the slightly more than three thousand-foot elevation at the Colorado-Kansas border. To the West the state is split east-from-west by the towering Continental Divide, the "Shining Mountains" that reach points more than fourteen thousand feet high. They are the source of six great river systems that water nineteen other states.

This land was known to primitive Folsom and Yuma people who migrated across the Bering Strait from Asia between ten thousand and twenty-five thousand years ago. In southwestern Colorado, the Pueblo culture flourished from about the beginning of the Christian era to approximately 1299. Late in the thirteenth century, the Ute and Apache came into the area. By the middle of the eighteenth century, the Comanches had arrived and had succeeded in driving the Northern Apaches south to the Rio Grande Valley. At this same time, the valley of the South Platte River was dominated by the powerful Pawnee people until they were fatally weakened by smallpox. By the time of the Stephen H. Long expedition of 1820, the Pawnee had moved to Nebraska, and the Kiowa, Cheyenne and Arapaho had become the dominant plains tribes. These two latter tribes were of Algonquian origins and had moved southwest from the regions of the Great Lakes. To the west, the Utes now controlled the passes into the mountains and much of the high country. South Park, for example, became one of their home territories.

Beginning in 1528, the first of many Hispanic people came north in a fruitless search for gold, becoming Colorado's first European explorers. When precious metals eluded them, the Spanish withdrew, concentrating their efforts south of the thirtieth parallel where gold was easier to acquire.

Traders, trappers, and an assortment of official American explorers carried word of the Pikes Peak country to the outside world, drew maps and gave names to many geographical features during the first part of the nineteenth century. But it was the Eastern American gold seekers who entered the land just after mid-century and were responsible for bringing great numbers of people to the West, inadvertently laying the foundation for another territory and ultimately for a new state.

At the time of the 1858 gold rush, there was no Colorado in the strict geographical sense of the word. The land that would become this state was still a part of four other territories. The Cherry Creek towns and most of the better front range mines were within the Kansas Territory, which had been created in 1854 under the Kansas-Nebraska Act. In 1858-59 the western boundary of Kansas' Arapahoe County was the Rocky Mountains. Future Breckenridge and its environs were a part of Utah at that time. Much of southern Colorado was within New Mexico, while Boulder and Gold Hill were a part of Nebraska. The fortieth parallel, today's Baseline Road through Boulder, separated Kansas from Nebraska. Future Colorado Territory was assembled from these sources.

The South Platte and the Arkansas are the two principal rivers that flow eastward as they drain Colorado's eastern plains. Along the northern portion, the broad North and South Platte branches come together near the Nebraska-Colorado state boundary. To the south the Arkansas emerges from the Front Range at Canon City, flowing eastward to the great Mississippi River. Between the Platte and the Arkansas, just north of the thirty-ninth parallel, stands a high seven thousand five hundred-foot headland that rises gradually above the eastern prairies. It is called the Cherry Creek Divide, since that stream originates there. It is also the source of an impressive number of north-flowing creeks, all of which eventually empty into the South Platte River. From the West, their names are Plum, Cherry, Boxelder, Kiowa, Wolf, Comanche, Bijou, Wilson, and Deer Trail. While most of these creeks join the South Platte at points further east, Cherry Creek's flow is northwest until it empties into the South Platte within contemporary Denver.

A few words about the Platte are appropriate here. Formerly it was an enormous river. In prehistoric times, it was about a mile wide where it passed through Denver. Its banks were, respectively, Federal Boulevard and Grant Street. As the earth grew warmer the volume of water decreased. Flood control projects, diversions and dams have further limited its flow. At the time of the gold rush, it was about twice its present size.

Cherry Creek is a name that almost replaced Pikes Peak as a designation for the gold fields of Western Kansas. Wild chokecherries have grown along its drainage from earliest times and were one of the fruits used by the Arapaho people in making pemmican. Colonel Henry Dodge's U.S. Dragoons found "cherries very plentiful" while camped along the drainage of that stream in 1835. When Col. John J. Abert's map of exploration appeared in 1845, the name of Cherry Creek appeared on it, suggesting that the designation was already a well accepted one.

But the origins of some other names are less clear, and geography could be confusing at times. Many names were of Spanish or Indian origin. Some others were French. Even in the first half of the nineteenth century, there were persistent rumors that the elusive yellow metal could be found along the streams that flowed out of the so-called "Shining Mountains," a not-too-specific name for the Stony or Mexican Mountains. Maps of the time were not very explicit, and the term Rocky Mountains had not yet come into general usage. As late as 1820, Stephen H. Long erroneously identified Longs Peak as Pikes Peak.

Likewise, the name of Pikes Peak was erroneously given to Colorado's 1859 gold rush despite the fact that the gold fields were located about eighty-five miles to the north. But Pikes

Peak had been "discovered" in 1806 and publicized after the War of 1812. Consequently, it became the only well-known landmark of importance and was mistakenly identified with the gold rush. Historian Francis Parkman refers to the mountain as Pikes Peak in his 1846 book, *The Oregon Trail*.

One of the best documented of the gold stories was reported in Zebulon Pike's own journals. When Lieutenant Pike was captured by the Spanish in 1807, he was detained for a time at Santa Fe. There he met mountain man James Purcell. Before coming west to become a trader, Purcell had been a carpenter at Bardstown, Kentucky. Purcell told Pike of having found placer gold near the headwaters of the South Platte River in South Park. Later, with the mountain man's typical disdain for material things, he threw the samples away.

In the 1830s a French Canadian known only as Duchet found gold in South Park while trapping on Horse Creek, a tributary of the South Platte River. Like Purcell, he too threw the gold away. Later, in Santa Fe, small particles of the gold were noted among the emptyings from his pouch. Interested observers asked Duchet to show them where he had found the nuggets. He tried to comply, but he was unable to locate the spot again.

In 1821 William Becknell, a trader out of Franklin, Missouri, became the person most responsible for development of the Santa Fe Trail. It was a businessman's road, built to connect the Missouri frontier with the New Mexican capital. Its northern branch followed the Arkansas River past Bent's Fort, turned south near Trinidad, and crossed the mountains by way of Raton Pass. It had been widely used by trappers and by the military in the years prior to the gold rush.

West of Bent's Fort, at Fountain Creek, many Argonauts left the Santa Fe road to avoid crossing the formidable barrier of the Rockies. They turned their wagons north, following along the Front Range shortcut to the well mapped Oregon-California Trail in Wyoming. Except for the few California-bound Forty-Niners who followed the Santa Fe Trail along the Arkansas River, Colorado was largely spared the wear and tear of the traffic emigrating to the West Coast. The Santa Fe Trail was the principal road to the West. It was the major thoroughfare, a much older route than the Oregon road.

Maj. William Gilpin, later to become Colorado's first territorial governor, was an ardent believer in the future of the American West. After exploring with Fremont in 1843, Gilpin reported gold finds along Cherry Creek and in several of its tributaries, in South Park, near Pikes Peak, along the Cache la Poudre River, and on Clear Creek. In January 1848, Gilpin was back in Colorado to pursue a band of allegedly hostile Utes in the San Juan Mountains. There he saw unmistakable evidence of gold and silver. By coincidence, this was the precise month and year during which gold was found in California. In the Sutter's Mill hysteria that ensued, Major Gilpin's accounts were hardly noticed. Gilpin was back again in 1853. This time he was traveling across the West as a member of Fremont's railroad seeking expedition. Gilpin, by now a true western enthusiast, was still fond of repeating the accounts of gold in the Rockies. Such was the reputation and status of the land that was about to be plunged into the turmoil of the Pikes Peak Gold Rush.

# 2.

# THE EARLIEST EARLY BIRDS

AS TIME PASSED, all of the older rumors of gold in the Rockies continued to flourish, and soon a variety of new ones began to appear. Several accounts were traceable to rather reliable sources. For instance, in 1852 soldiers from Fort Leavenworth under a Colonel Sanborn were pursuing hostile Comanche Indians in eastern Colorado. They followed their quarry along the Santa Fe Trail as far as the cut off to Fort Pueblo, then turned north toward Manitou Springs. Continuing north, they crossed the Palmer Lake Divide to the head of Cherry Creek and followed its course down to the South Platte River. Along the way, they were treated to mountain men's tales of gold. Upon reaching Fort Laramie, they repeated what they had heard, probably with some semi-official embellishments.

Capt. Randolph B. Marcy was first exposed to the gold seeker's fever on May 3, 1858 while he was on the way back to Fort Bridger, Utah, from New Mexico with supplies and troops. Along the South Platte River, at the mouth of Cherry Creek, he met the redoubtable farmer, Charlie Autobees. While awaiting the construction of a river raft, a civilian with Marcy's party washed a minute quantity of low grade gold from a sandbar. Back at his home in St. Louis, the civilian became fairly noisy about his good fortune, showing the glittering samples to any of the unemployed hangers-on who cared to look. Another eye witness story of gold in the Rockies was told by a Delaware Indian named Fall Leaf, who had worked with Col. Edwin Sumner's expedition as a guide. Sumner's party was a military one that had been sent out to discourage the Cheyenne from raiding along the South Platte River. Upon returning to

Kansas, Fall Leaf told about having seen shiny gold nuggets in the bed of a mountain stream. The loquacious Delaware repeated his story to John Easter of Independence, who ran a butchering operation in the nearby town of Lawrence.

Fall Leaf made his home with a small band of the Delaware tribe that occupied lands along the north side of the Kansas River. Occasionally he and other Indians sold cattle to butcher Easter. Fall Leaf carried his gold nuggets tied up in a handerchief. While trading in Easter's shop one day he displayed the shiny particles and told Easter that he had found them while in the field with Colonel Sumner's expedition. He described stopping at a mountain stream to water his horse and to get a drink for himself. The gleaming elements lay among the gravel in the stream bed.

Almost at once, Easter put together a party to investigate the validity of the Delaware's story. In order to assure the availability of Fall Leaf as a guide, Easter offered to provide the Indian's family with rations for the estimated half a year that the party would be gone. But suddenly, Fall Leaf seemed to lose his enthusiasm for travel. His excuses for not going included a fear that the party was too small and that he could no longer recall where he had picked up the nuggets. A third reason was his alleged ill health, brought on by an injury he had sustained in a barroom fight. But newsman Henry Villard had another version, Villard said that Fall Leaf's tribesmen had threatened to kill him if he revealed the location to white men. Allegedly, they were furious that Fall Leaf had disclosed the story to Easter.

Whether or not Fall Leaf accompanied the Easter party was academic at that point. Enthusiasm had been whipped to a frenzy and they waited only for spring and clearer weather. Easter's party consisted of 11 ox-drawn wagons and about fifty people. Another source differs slightly on the group's make-up, insisting that it numbered forty-six men, two women, and a child. One of the women was Julia Holmes, who endured much ridicule for her vigorous and noisy advocacy of women's rights.

To demonstrate her durability and equality, Mrs. Holmes

persisted in walking about ten miles each day beside the family's ox-drawn wagon. At night she insisted on taking her turn at standing guard along the trail. Furthermore, she wore "the bloomers." Her endurance marathon continued for the five weeks that were required to cross the plains, Julia claimed to be the first of her sex to negotiate the prairie while wearing her suffragette bloomer costume. At the great bend of the Arkansas River, friendly Arapaho Indians allegedly offered to swap both ten ponies and some buffalo hides for Mrs. Holmes. Her appearance may have accounted for the glib and often repeated exclamation of the buck who sought to purchase her. He supposedly looked at her and whooped, "Heap squaw! Whoa! Heap God damn!"

Some accounts identify this Kansas group as the Easter party while others refer to it as the Lawrence party. Their route west was predictable. They followed the Santa Fe Trail across Kansas to the Arkansas River, then north toward Pikes Peak. From the Missouri River to the diggings involved a distance of just over six hundred miles. North of the Arkansas River, the Easter party ascended Fountain Creek to the future site of Colorado Springs.

Early in July they camped for a few days near the red sandstone monoliths of the Garden of the Gods. From this point, two small groups headed up over Ute Pass to try their luck in South Park. While the main party awaited their return, three adventuresome men decided to use this resting time for a climb of Pikes Peak. Upon their returning to camp, Mrs. Holmes announced that she could handle anything that men could accomplish.

And so, early in August, Julia and her long suffering husband, James, set out to conquer America's best known mountain. Their packs were filled with bedding, bread, and a copy of Emerson's essays. They camped in the timber the first night. To the east they looked out across the plains toward Kansas. Below them the canvas-topped wagons of the Easter camp were visible. Two days later, above timberline, they experienced a brief snow flurry. Near the summit, they were snowed on a

*From the Collection of Evelyn and Robert L. Brown*

Although far from the actual gold discoveries, the name of Pikes Peak became synonymous with the gold rush.

*From the Collection of Evelyn and Robert L. Brown*

"For purple mountain majesty above the fruited plain," the panorama from the summit of Pikes Peak that inspired Julia Holmes and Katharine Lee Bates.

second time and the weather became bitterly cold. They wrote letters to friends, read a passage from Emerson aloud, and when the weather cleared they paused to view the majestic ranges to the west.

Julia and James camped only one night during their descent. Shortly after noon, they reached Easter's campsite. Julia Holmes had become the first woman to conquer Pikes Peak. Her remarkable accomplishment was later set down in a book called *A Bloomer Girl on Pikes Peak.*

Southeast of their campsite, the Easter group laid out the building sites for a new town. They called it El Paso. The location chosen was within the limits of contemporary Colorado Springs. Late in August they contemplated moving to the San Luis Valley and also discussed spending the winter in New Mexico. But fate intervened with the news that gold had been found by Green Russell. So they reversed directions, heading north into the pandemonium at Cherry Creek. Their lots at El Paso were sold for a mere two thousand dollars. Incidentally, the Holmes family finally gave up on gold seeking. For many years they lived in New Mexico. Later Julia became the first woman to secure a position with the Department of Education. Finally she divorced husband James and became an active suffragette.

By September the Lawrence-Easter party had arrived at its latest location. Their claims along the South Platte River were close to the recently abandoned Green Russell placer. The site they chose was on Dry Creek and was part of the winter campground of the southern Arapahos.

Once more they were busily laying out a townsite. Like El Paso, this was to be another "first town." It was called Montana City, the initial organized community on the site of what was to become Denver. Montana City's town company was created on September 7, 1858. Although it was the first, it lasted for less than a year. Easter's people soon found Russell's abandoned diggings and quickly set up their own operation. Wood from their wagon bodies was used to construct sluice boxes. Their plan was to spend the winter months at the site.

With regard to contemporary landmarks the Montana City townsite is located to the south of West Evans Avenue, east of the South Platte River, extending to a vague limit somewhere beyond the Santa Fe Railroad tracks. Montana City's southern extremity was near the confluence of Little Dry Creek and the South Platte River. Since no formal boundaries were ever adopted, all of the aforementioned limits are approximations.

But the Easter people soon discovered that most of the action was occurring about ten miles north, at the point where Cherry Creek empties into the South Platte River. Here, too, was the terminus of most of the emigrant trails from the east. So they packed up their wagons again, moved north, and started a third town. It was called St. Charles.

A survey of their new townsite began almost at once, but only one cabin was ever built. In late September the weather turned cold. On October 2, the Kansans departed for home, hoping to spend the winter in Lawrence planning the future of their newest venture. Ironically, their optimistic dreams of the future were about to be shattered. Still another group of Kansans was on the way to Cherry Creek. This was the opportunistic and aggressive William Larimer party from Lecompton. St. Charles, unlike Montana City, was destined to play an important role in the gold rush saga.

Inevitably, in the depression ridden economy of the Missouri Valley, news of anything such as a gold find spread like the proverbial measles in a kindergarten. Frontier journalists copied each others' stories, embellishing upon the few scanty facts with each rewrite. Consequently, when spring came in 1859, several well-equipped parties of modest size had been assembled in western Missouri. Other prospecting ventures were organized at Bellvue, Omaha, Plattsmouth, and Florence, Nebraska, and at Council Bluffs, Iowa.

# 3.

# THE CHEROKEES

IF ONE DISCOUNTS RUMORS that minute quantities of placer gold were found along stream beds in the Massachusetts Bay Colony, the next find of substance occurred in 1799 when gold was discovered on John Reed's farm in North Carolina. Reed found a nugget in 1802 that he sold for $350. A later appraisal placed its value at $8,000. Farmer Reed sued, but got only $1,000. Later, gold worth $10 million came from the Reed farm. One nugget weighed twenty-eight pounds.

Then came the gold discovery that was made in western Georgia in 1827. Typically, the Georgia find was made in mountainous country near the southern tip of the Appalachian chain. Auraria, a tiny community whose Latin name means gold, became the center of the southern mining region. Before it was over, some twenty million dollars had been taken from the Georgia gold belt.

Since much of the land surrounding the discovery sites was home to the Cherokee people, ways were devised to assure transfer of the land to Anglo-Saxon control. The Cherokees, therefore, assumed a place among the earliest of a long succession of native Americans who found themselves dispossessed in the process of our country's voracious race toward Manifest Destiny. In this instance, the Cherokee were moved to the Indian Territory within present Oklahoma. Since many in the Georgia tribe were reluctant to accept this as their just fate, when gold was discovered in California during January of 1848, a party of Georgia-Oklahoma Indians talked it over and decided to head west.

Because many of the Indians had gained experience while

On the Santa Fe Trail, Bent's Fort was reconstructed in 1976 by the National Park Service.

In the center foreground, the low swale is the original Santa Fe Trail.

working in the Georgia mines prior to expulsion, there seems to have been some vague notion of striking it rich in the Pacific coast fields. With gold in their pockets, legal talent could be retained to sue for the return of their ancestral homeland. Lewis Ralston and his brother, Samuel, whose names now grace several Colorado locations, encouraged a number of educated Cherokees from north Georgia to join the venture.

A casual check of any national map will indicate that the most practical route from Indian Territory to California would pursue a more or less northwest direction. So, the optimistic little band of people henceforth known to history as the Ralston party, followed the already well-established Santa Fe Trail. This approach carried them into what is now southern Colorado.

Near Bent's Old Fort, they left the established trail, turned northwest past Pikes Peak, following an older trail of probable Indian or Spanish origin. It continued northward paralleling the front range of the Rockies. With continued use, it became known as the Old Cherokee Trail. Much later, its northern part was used during the Civil War years to bypass Indian troubles along the better-known Oregon Trail.

On June 22 the Ralston party camped north of the future site of Denver and found small amounts of gold in that tributary of Clear Creek that is still known as Ralston Creek. It was gold, assuredly, but it was not present in a quantity that would justify their remaining to develop a claim. Continuing north, they found gold studded quartz rock beside the Cache La Poudre River, but only in a very limited quantity. At last they came to the California-Oregon Trail and followed it westward to their destination. Despite their experiences in the Georgia goldfields, they were unrewarded in California. Late in the spring of 1857, they turned eastward toward home.

Autumn of that year found them retracing their steps along the front range of the Rockies. Some of them remembered what they had found here seven years before. They also noted a general similarity between this country and the gold bearing regions of California and wondered if the resemblance could be more than a coincidence. The decision was made to delay

*From the Collection of Evelyn and Robert L. Brown*

The Old Cherokee Trail's scars run diagonally across this picture, while tepee rings are visible on either side of the ruts.

*From the Collection of Evelyn and Robert L. Brown*

In Jefferson County the route of the Apex and Gregory Toll Road can now be hiked.

a bit while they tried their luck. Later that summer, upon their arrival at Westport, Missouri, they told of having found minute traces of placer gold in Cherry Creek. Although their discovery was more exciting than substantial, a plan for future exploration was already evolving with the group.

When they reached the Missouri River, the party split. Some of them returned to Georgia, while others headed for their homes in the Indian Territory. That winter the Cherokees showed their gold samples around and talked openly of their find. John Beck, a Cherokee Baptist preacher, had been one of the original members of the California expedition. Like the others he had found no gold there. He spent the early winter months of 1857 planning for a party of his own to seek out the yellow metal in the Rockies. Beck wrote to a relative in Georgia urging formation of a second party there. His hope was that the group might follow Ralston Creek to its beginning and locate the source of the gold.

Beck's letter went to a relative by marriage, William Green Russell, another veteran of the unsuccessful trek to the West Coast. Another source tells quite a different story about the California escapade. According to Emma Dill Russell Spencer, a direct descendant, Russell and his brothers found enough gold in California to pay for brother Levi's Philadelphia medical school tuition, and to enable Green and brother Oliver to purchase a Georgia plantation. In any case, the three Russells could not quite forget what they had seen in the Pikes Peak country; they were anxious to return.

Green Russell, in the unfeeling vernacular of the times, was known as a squaw man because his wife was of Cherokee heritage. The Russell family lived in Dahlonega, Georgia, a town located only five miles from gold rich Auraria. In both Georgia and Oklahoma, the cold months were passed in preparation for an early departure to the Pikes Peak country. The Georgia group left first. Those in the Cherokee Nation did not depart until May. Their plan called for a merger near the Great Bend of the Arkansas River. Actually, they joined forces on June 2 at a point just west of Larned, Kansas. The final count showed one hundred four people in fourteen wagons. In addition to nineteen men from Georgia and forty-six Cherokees from Indian Territory, two parties from Missouri joined the group. Historians have referred to this assemblage as both the Russell and as the Russell-Cherokee party. Their trip west took more time because Preacher Beck, in common with the earlier Mormon migrants, insisted that the party not travel on Sundays.

Their route to Colorado was virtually identical to that followed by the original Cherokee prospectors of 1850. As they approached their destination, Russell paused on the Cherry Creek Divide and placered out a few traces of gold. At a point about four miles north of present Franktown, they established a tiny mining camp of brief duration called Russellville. By mid-June of 1858, there were nine log houses clustered around the site of their placer. A large barn, a stockade, a corral and a log tavern completed the roster of structures. Probably its out-of-the-way location doomed Russellville. The more direct Smoky Hill route avoided the divide and provided a far better access to Denver. No trace of the old Russellville remains within what is still called Russellville Gulch.

Continuing on, the Russell party arrived at its destination late in June of 1858. After ten days of difficult living and working conditions that yielded almost no gold, John Beck quit the prospecting game and led his contingent of Cherokees back home to Indian Territory. A few of Russell's Georgians left with Beck. But Russell himself and a few diehard friends elected to remain on the South Platte River to continue prospecting.

During the second week of July, their persistence was rewarded. Their strike was near present Englewood in a small pocket that had been created where Dry Creek empties into the South Platte River. In their journal, the Russell brothers referred to this site as Placer Camp. Although the gold was of good quality and was thought at first to be a bonanza, its quantity was limited. Russell took a few hundred dollars from the deep pocket. By mid-August, its value was exhausted for everyone except those people who now used the discovery to spread the story across the economically-depressed Missouri Valley.

Next to come on the scene was John Cantrell, a mountain trader from the Salt Lake Valley. Cantrell panned out a meager three ounces of gold from the Russell location. When he reached Kansas City, Cantrell showed off his find. He spread the story that Green Russell's party had made a thousand dollars in ten days of panning. The *Kansas City Journal of Commerce,* in its issue of August 26, 1858, ran the story. Three days later the *St. Louis Republican* reprinted it. On August 30 it was carried almost verbatim in the *Daily Journal* at far away Boston. Needless to say, the story grew richer with each embellishment. During the remainder of 1858, several midwestern newspapers went to press with optimistic reports of gold discoveries in the western part of Kansas Territory. Among them were the *Weekly Kansas Herald,* the *Missouri Republican* at St. Louis, and the *St. Joseph Gazette.*

*From the Collection of Evelyn and Robert L. Brown*
The original Ute Trail was the route from Colorado City to the upper Arkansas Valley.

William Green Russell of Georgia found placer gold near Denver and in Gilpin County.

Green Russell himself was one of the first to become alarmed at the gross exaggerations being printed about the "Russell Bonanza." So he boarded his wagon and set off for the East, telling people he met that his find had been "modest" and that the pocket now contained no more gold. But in an environment where the aftereffects of the Panic of 1857 were still being felt, the westbound Argonauts mistook his altruism for greed and continued on toward Cherry Creek.

As Russell drove further eastward through Missouri, Tennessee, and into Georgia, the rumor that he had found gold deposits in the Rocky Mountains continued to follow him, still growing with each telling. His constant protestations served only to supercharge the gossip. Within weeks of his return home, stories of his supposed great wealth were reprinted up and down the East Coast. Each newspaper tried to outdo its competition with ever more extravagent reports.

Poor Green Russell became a creature of history. Warnings aside, he could not curb the speculation. A great hunt for quick riches was on. Green Russell, an honest man, played no further role until he returned to Colorado again in May of the following year, 1859.

# 4.

# JACKSON AND GREGORY

ALTHOUGH THE PLACER GOLD found by the Green Russell party had provided the initial spark that started the gold rush, it was the substantial discoveries of two other men that gave substance to the migration. Together, they saved the entire event from becoming a monumental fiasco. Both of these better finds occurred early in 1859. They also were made by southerners, George A. Jackson and John H. Gregory. The Jackson discovery came first.

Jackson was born on July 25, 1836 at Glasgow in southern Missouri. His son-in-law, Mark W. Atkins, once stated that his birth date was July 24, 1832. His daughter said that he left for California when he was only sixteen years old. There he tried and failed in the Sierra gold fields. While still on the coast, he tried farming, but gave that up, too, and began the return home. He apparently took his time crossing the northern Rockies, spending a total of six months just getting through them during the summer of 1857. After reaching Missouri, he returned to the West and spent time in the Sweetwater country of Wyoming. Summer of 1858 found him dallying with three old mountaineers near Fort Laramie when word came of the gold excitement around Cherry Creek.

At Fort Laramie, Jackson was subjected to all the exaggerated accounts of Green Russell's strike some one hundred eighty miles to the southwest. He put together a prospecting party of twenty men. Moving south, they washed most of the streams as they worked their way down the front range. In company with three mountain men whose names have not survived, Jackson investigated sandbars along an eighty-mile stretch of

George A. Jackson found gold at the future site of Idaho Springs.

John H. Gregory of Georgia found the first lode gold in Colorado.

*Collection of Freda and Francis Rizzari*

Idaho Springs grew up around the George Jackson discovery. The mill at the right is the Argo.

*From the Collection of Evelyn and Robert L. Brown*

Idaho Springs still contains many original structures. Inset shows bar tokens from the town.

Here was Idaho Springs as photographed by William Henry Jackson.

Modern Idaho Springs is a busy mountain community.

the Laramie River. But they raised no color. Finally, with resources almost exhausted, the Jackson party arrived in the Pikes
Peak country. Minute quantities of placer gold were found near
the point where they crossed St. Vrain Creek. By this time,
the chill of late autumn made gold seeking less comfortable.
A few days later, the first snow fell, hastening the decision to
establish their winter camp. Jackson was now just twenty-four
years old.

Jackson's group chose to hole up at Arapahoe City, on present
West Forty-Eighth Avenue, west of contemporary Denver. Here
they spent the early months of winter in the traditional pursuits
of idle miners everywhere. They speculated in town lots, gam-
bled, slept, whittled, fought and drank to excess.

But Colorado's winters are lengthy, and Jackson succumbed
to "cabin fever." With two companions named Tom Golden
and James Sanders, he started into the snow-covered mountains
above the Clear Creek Valley on the day after Christmas, 1858.
The men separated, probably to hunt for elk or deer. Jackson
moved up Mount Vernon Canyon, possibly crossing Bergen
Park. From there he started to climb Squaw Pass. Before reaching the summit, he dropped down Soda Creek into the valley
of South Clear Creek. Far below, he saw what he assumed to
be smoke from an Indian cooking fire, a possible source of a
free meal. But instead of smoke, Jackson found only rising
vapors from the mineral hot spring that gave the later town
of Idaho Springs its name.

On January 5, Jackson camped about a half-mile west of
the spring where today's West Chicago Creek empties into
Clear Creek, then called the Vasquez Fork. While melting ice
for coffee water, he happened to glance down through the frozen
creek surface. What Jackson had chanced upon was a rich deposit
of placer gold, washed into the natural pocket created by cen-
turies of turbulence at the junction of the two streams.

One tradition asserts that Jackson chose this particular spot
as a place where he could get out of the wind and build a
cooking fire. Another version tells us that Jackson paused there
to rest his dog, Drum, recently lamed in a fight with a carcajou

*Collection of Freda and Francis Rizzari*

In the early 1860s, William Gunnison Chamberlain made this rare photograph of two water powered arrastras in operation at the town of Montgomery in South Park.

*From the Collection of Evelyn and Robert L. Brown*

Evelyn Brown and pet pose beside an old arrastra, minus the headframe.

(wolverine). In any case, he spent four days at the site of his discovery, carefully noting the lay of the land so that it could be found again. Back at Arapahoe City, he kept his own counsel for several weeks. Later he confided in Tom Golden, whose mouth was said to have been "as tight as a #2 beaver trap."

On March 17 a prospecting party from Chicago came through Arapahoe City with a supply of food. Possibly because of the nature of the supplies he was stockpiling, the suspicion was soon rampant that Jackson had located a bonanza. After much importunate pleading, the Chicago men won Jackson's confidence and were allowed to accompany him to the placer site. In another version of the story, it was Jackson who took the initiative and was finally able to interest the Chicago men in helping to develop his find.

With his new partners, Jackson returned to the site early in March. Their supplies were hauled up in a wagon. So rough was the going that they sometimes took the wagon apart, carrying the pieces around the worst places. As partners, they staked out several claims along the stream bed. From the circumstances of this association with the home town of these men, the stream is still called West Chicago Creek.

By mutual agreement, the partners decided to continue trying to keep their good fortune a secret. But in early May Jackson came down from the mountains to purchase more supplies. Foolishly, he paid for them in gold dust. Word of his impropriety spread rapidly and a gold-hungry throng followed at a discreet distance as he made his way back to West Chicago Creek. The golden cat was out of the bag. Sluice boxes soon lined the creek beds in both directions. Noisy Spanish-style arrastras were in use and a settlement called Spanish Bar had begun to grow up near Jackson's diggings.

Meanwhile, not far away, another substantial discovery was made during that same month of May. In total value and in social significance, it would soon eclipse the meager Jackson placer, giving substance to the entire Pikes Peak Gold Rush.

John H. Gregory, like Green Russell, was a Georgia native. He had first learned about digging gold while working in the

Mountain City grew up around John Gregory's gold mine. Some of Central City's buildings are visible in the middle distance.

Only the scars of early streets and the background mountains identify the location of Mountain City.

Dahlonega County mines of southern Georgia. Physically, Gregory has been described as sandy haired or red headed. His vocabulary was coarse and colorful in the extreme. Every other word was an oath. In common with many of his contemporaries, Gregory had lost nearly everything in the Panic of 1857. Two years earlier, in 1855, gold had been found on the Fraser River in New Caledonia, present British Columbia. A modest rush of Americans took the sea route north from San Francisco in 1858. One of the stampeders was John Gregory. A compulsive but poor would-be prospector, he headed overland toward the boom town of Barkerville in the Cariboo mining district.

Following the Oregon Trail, Gregory both summered and wintered at Fort Laramie. For some unknown reason, his grand plan of journeying to British Columbia was forgotten. Earlier, he had worked in the West as both a trapper and a hunter. At Fort Laramie, he was employed as a teamster. One other source is at variance on this point, insisting that Gregory worked only as a common laborer, a not very important detail. He may have been herding bull teams when word came of the gold discoveries near Cherry Creek. Almost at once, he quit the job and headed south. Gregory followed the Cache La Poudre River southeast before moving down the front range.

Another account tells how Gregory left Fort Laramie with a government wagon train bound for Fort Union in the New Mexico Territory. He was nearly destitute of both provisions and clothing. Leaving the train, he spent about a month lounging about at Arapahoe City, listening to the tantalizing rumors about Jackson's find of four months before. At Arapahoe City he lived off the generosity of Capt. Richard Sopris until told to "move on." He checked out several streams, possibly as far south as Pikes Peak itself. Finally, he chose the valley of Clear Creek, turning west along its rocky watercourse. With a companion named William Kendall, he started up the valley. Two white mules carried their provisions. Being an enthusiastic prospector, Gregory stopped periodically to pan the sandbars for signs of gold.

Fourteen miles above the settlement, he arrived at the point where North and South Clear Creeks merge their waters. Here Gregory was truly in a dilemma. Which way should he go to reach Jackson's rumored diggings? So he panned both streams and chose the North Fork on the basis of the more abundant gold particles he was able to take from that branch. Using the same technique, Gregory continued up the canyon, eventually entering the gulch that now bears his name. At Prosser Gulch, a half mile up, he turned again, following a lead to the first lode gold discovery in the territory. An unseasonal April snow dictated a retreat from the mountains.

At nearby Golden Gate City, Gregory was able to interest businessman David K. Wall in supplying provisions. A party was formed, although few of its members believed Gregory's report. Among the members were Wilkes Defrees and his brother, Archibald, Dr. Joseph Casto, James Wood, H. P. A. Smith, C. H. Butler, James Hunter, C. Dean, Capt. W. H. Bates, Charles Tascher, and William Ziegler of Missouri. This was the group that accompanied the discoverer on the return trip to the site of his find. Two yokes of oxen and a number of pack animals carried their provisions.

For this second trek, a different route was followed. They made their way up Golden Gate Canyon to its head, but did not go by way of the spine of the canyon. Using the higher ground, they crossed Guy Gulch to Ralston Creek, following it to Dory Hill before dropping down toward Clear Creek. Their trek consumed four days. On May 6, 1859 they dug through an accumulation of ice and snow and took four dollars worth of gold from their very first pan. Gregory was stunned by what he had found.

"By God, now my wife can be a lady!" he said. "My children will be schooled!"

On June 24, William Newton Byers, founder and editor of the *Rocky Mountain News,* arrived in the gulch to interview Gregory. By this time Gregory was delirious from loss of sleep. The crudely retorted gold was hidden beneath an overturned cast iron fry pan. Gregory was seated on the pan with a rifle laid across his knees to discourage theft. Byers took pity on him. In company with an extra driver, the editor relayed the gold through to Omaha in a mere twelve days. The Gregory samples were placed on public view in the window of an Omaha bank. With this ostentatious display, the last doubts vanished. Now it remained for the rumor mongers to interpret the gold's value and to spread the story eastward.

Back in his gulch, Gregory sold his mine for twenty-one thousand dollars and went about cashing in on his reputation. He hired out to prospect for others at two hundred dollars a day. Many of his locations became fine producing properties, gold being so prevalent he couldn't miss. Each such instance enhanced his reputation as a expert. Wisely, Gregory decided it was time to make his way back to his Georgia home. He never returned to the West. It was the Gregory find that gave substance to the Pikes Peak Gold Rush.

News that Gregory had found lode gold filtered quickly down from the mountains. From Arapahoe City, it spread eastward to Auraria and Denver City. When summer turned to fall, some fifteen thousand men were settling into the tiny confines of what was already being called Gregory Gulch.

As news of the Gregory and Jackson finds reached the Cherry Creek towns, it virually emptied those settlements. Would-be prospectors stampeded up every side canyon in the front range. Only the saloon keepers and the bums who followed them were left behind. By the end of June of 1859, another ten thousand optimistic gold seekers had come to the North Fork of Clear Creek to try their luck. Others set up their sluice boxes along the lower creek beds, up the gulches, and on the surrounding hillsides.

A thick dust pall hung over the valley each day as the miner's blasting and digging activities tore up the terrain. Some eighty-five million dollars, mostly in gold, was eventually excavated from the Gregory district. Understandably, there were many visitors. Even New York editor Horace Greeley made a personal pilgrimage into the West to check the facts before publication in his newspaper. After his June visit to the Gregory district, Greeley wrote, "About a hundred cabins are being

Black Hawk, at the lower end of Gregory Gulch, milled most of Gilpin County's ores.

Black Hawk, as it appears today, contains many original buildings. Inset shows bar tokens from a few of the town's many saloons.

This view of Central City shows St. Aloysisus Academy, top, and the Catholic Church at right.

From the same angle, here is a contemporary view of Central City.

This early view of Central City looks north across the town.

From the same angle, here is Central City today. Inset shows early saloon tokens.

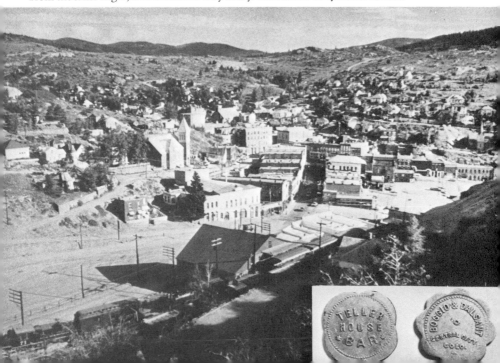

built and three or four hundred more are in immediate contemplation. As yet," he added, "the entire population of the valley sleeps in tents, or under booths of pine boughs, cooking and eating in the open air. I doubt there is yet a table or a chair in these diggings, eating being done around a cloth spread on the ground, while each one sits or reclines on mother earth."

Other men came, too, men who would never swing a pick nor wet a pan in an icy mountain stream. The Clear Creek towns proved to be very attractive to a wide variety of confidence men, mining speculators, and town boomers. Lawyers did well, too, considering the vast number of disputed claims. Whiskey sellers' wagons not only lined the emigrant's roads, but frequently were "expanded" to become the first businesses in the new towns. All that was needed to enter the saloon business was a tent, two whiskey barrels, a rough plank, one or two paintings, and some tin cups. "Taos Lightning," a crude corn whiskey, was dispensed at fifty cents a shot. Its curious name came from the fact that it was fermented in nearby Taos, New Mexico. Among its devotees was a widely accepted belief that when one downed two cups of the brew, what followed was a sensation comparable to being struck by a bolt of lightning.

One visiting journalist described the network of ramshackle towns as resembling the rungs of a ladder, blending into one another along the steep three miles of the Gregory district. At the foot of the hill was Black Hawk, the milling and refining town, where smelters lined both sides of North Clear Creek. Above it were Gregory Point and Mountain City, two towns that had grown up around the original Gregory mine. Richard Sopris had established Mountain City in May. By mid-June, a visitor observed that it contained more than one hundred miners' dwellings, clinging doubtfully to the sharp southern side of the gulch. Stores appeared, also hotels, saloons, gambling houses, an express office, and a printing shop. Physicians and lawyers hung up freshly-lettered signs as town life came to the mining district.

Continuing up the hill, one came next to Central City, which became the largest of the towns and the unofficial capital

Collection of Fred and Jo Mazzulla

This was Nevadaville, the Irish-Cornish town above Central City.

From the Collection of Evelyn and Robert L. Brown

The largest structure was Nevadaville's Masonic Hall. Inset shows bar tokens from
Nevadaville saloons, including rare Bald Mountain imprint.

of the "Little Kingdom of Gilpin." Near the head of Eureka Street was the tiny hamlet called Eureka, located just east of the three cemeteries, Missouri City was situated a short distance southwest of Central City. With discovery of the Burroughs lode, Nevadaville, a predominantly Irish town, grew up at a point barely a mile above the softball park, now a parking lot. Because of the profusion of similarly-named towns in the West, the postal authorities refused to accept the Nevadaville name. Hence the post office there was always known as Bald Mountain. Tenaciously, the residents refused to use the new name, except for one lone saloon keeper who displayed the unpopular Bald Mountain name on his bar token.

Here the Irish majority lived, clannishly together in an incessant turmoil, dealing needful cuffs among their screaming children. Adding to the turmoil, Nevadaville boasted of its saloons, as well as an abundant number of rich mines, like the Casey, on adjacent Quartz Hill.

Cornishmen, "Cousin Jacks," the world's finest miners, were brought in to work in the deep mines around Nevadaville. Probably because of the competition for jobs and cultural differences, a Cornish-Irish feud broke out and continued for many years. Cousin Jacks were staid, substantial men with florid faces and sandy hair, addicted to broad humor which showed itself in rough practical jokes and horseplay. At night, behind curtained windows, their voices were heard rising and falling in the measured cadences of sad Welsh ballads.

In Spring Gulch stood a town called Springfield. Along the Virginia Canyon Road were Bortonsburg and Russell Gulch, site of Green Russell's later gold placer. Russell Gulch, a fairly large town where about one thousand people settled, was less than two miles southwest of Central City. In a single week there, miners took out placer gold worth nearly $2,000. Later, the Russell Gulch mines produced $35,000 weekly.

When the Civil War began, the Russell brothers sold their Colorado holdings and slipped quietly away. Upon reaching Georgia, Green raised a regiment and fought for the Confederacy. He returned to Colorado for a brief visit in 1872. Russell

*Gilpin County Historical Society*

This early view of Russell Gulch shows the town from the cemetery road. Inset
shows Russell Gulch saloon tokens.

*From the Collection of Evelyn and Robert L. Brown*

Much of old Russell Gulch may still be seen.

Gilson Gulch was a tiny camp between Russell Gulch and Idaho Springs.

Here is a contemporary view of the Gilson Gulch site.

died on August 14, 1877 in Briarwood, Oklahoma, and was buried in the town schoolyard.

Farther down Russell Gulch, on the original Virginia Canyon road, was Gilson Gulch, a much smaller town with houses arrayed along a single crooked street that conformed to the terrain's contour. Above it on the steep sides of the gulch were its mines — the Sun, the Moon, and the Gem Consolidated. Beyond it, two twising roads run down the canyon into Idaho Springs.

# 5.

# WEST TO THE ROCKIES

THE SIMPLE EXPEDIENT of time often has a wondrous effect on human events. Out-of-work farmers and other midwestern males spent the winter of 1858–59 discussing the gold strike stories that had found their way east. With each retelling the prospects continued to grow. By spring, many people had decided to cross the prairies to see for themselves. During a single week in May 1859, nearly six hundred wagons departed from Omaha. America's Panic of 1857 had depressed the economy, caused massive unemployment, and had pushed people to embark on foolhardy ventures that would never have been considered in more normal times.

Factories closed their doors, banks failed, and thousands of easterners found pink slips in their pay envelopes. Under the circumstances, out-of-work easterners literally lost their heads over gold dust prospects. They rushed headlong toward far-away places of which they knew nothing beyond the nebulous rumor that precious metals had been found there.

The presence of even a little gold in Colorado was the worst kept secret of the decade. When it became generally known, the rush that followed evolved into a powerful social force that moved thousands of depression-ridden Americans from the East and Midwest. They traveled to an unsettled land which, except for the promise of quick riches, would have been left for at least another decade to a few hardy trappers and Indians.

With hard times, even the words "Pikes Peak" became exciting, carrying with them the connotation of quick wealth. These two words became the enticing foundation of hopes and dreams of a better way of life. Consequently, the Pikes Peak

*From the Collection of Evelyn and Robert L. Brown*

On Deep Rut Hill, the wheel scars of the Oregon Trail have cut deeply into the sandstone.

*Western History Department, Denver Public Library*

The wagon adjacent to this trailside saloon displays the familiar "Pikes Peak or Bust" logo.

Gold Rush played a vital role in peopling and civilizing the vast plains area immediately to the east of the Rocky Mountains, and finally filling up even the mountains themselves.

Conservatively, at least fifty thousand people, mostly men and primarily from the East and the Missouri Valley, made up their minds to emigrate in the spring of 1859. They liquidated or sometimes just abandoned their assets at record speeds, and managed to find transportation to the gold fields. By the end of the summer of 1859, it was estimated that one hundred fifty thousand had crossed the Great Plains.

Most of the "Peakers" were comparitively young. Few were over thirty. The greatest numbers came from New England, Georgia and Virginia. Beyond that, listed in declining order, others came from Ohio, New York, Illinois, Missouri, Indiana, and Michigan. Among foreign born, Canadians led in frequency, followed by the Irish and Germans. Most of them were out-of-work farmers, although many listed their occupations as "miners." Among the more honest were those who claimed to be laborers, carpenters, farmers, teamsters, and clerks.

Some rode mules or horses, others walked, pushing a wheelbarrow that contained their belongings. A few followed the example set by the Mormons in the 1840s, pushing and pulling handcarts that were laden with a variety of goods and mining implements. Many of the handcarts were hauled by women. One was seen being pulled by five young girls in bloomer costumes, laughing and singing as they plodded along.

Some foot travelers brought carpetbags suspended from the ends of poles or from shovels carried over their shoulders. In one party there was a buggy loaded with mining tools. It was pulled by eleven men, led by a young male dressed in a fine cloth coat, silk stovepipe hat, and patent leather boots. Another top-hatted dude was seen leaving Westport in a fringed buggy pulled by two strong men.

When Argonauts responded to newspaper advertising for the hastily organized transportation companies that offered transportation to Pikes Peak, they often found that "passage" meant transportation for their baggage only, while they were expected to walk the entire seven hundred miles. One young man was seen pushing a wheelbarrow piled high with flour sacks to shovel his gold into. A few light wagons crossed the prairies drawn by large dogs. One Wyoming prospector used a small wagon pulled by two Newfoundland dogs, two pointers and a pair of greyhounds.

The excitement had barely begun when modest numbers of gold seekers became convinced that they had been misled by inaccurate reports of easily acquired riches. They in turn spread panic among their fellows until the number who had opted to return home nearly equalled those still moving west. At night prairie campfires were visible in an east-west panorama that sometimes stretched from horizon to horizon. Some were back home within six months of their departures, labeling the whole thing a "hoax or a humbug."

Of the estimated one hundred fifty thousand who headed west in 1858–59, at least a third turned back before coming in sight of the mountains. Another twenty-five thousand returned east when they found the work of gold extraction harder than described in hometown newspapers. In all, about thirty thousand stayed to form the population nucleus in the new land. One curious side effect of this eastward movement was the fact that the "go backs" took blue spruce seedlings east with them. Today these trees, formerly native to the Rockies, may be seen all over the East.

In those days before bumper stickers, it became chic to paint the words "Pikes Peak or Bust" in large block letters across canvas wagon covers. Many hundreds of such vehicles were seen and were duly reported in dairies and in newspaper stories. But curiously, few actual photographs have survived showing this logo emblazened across the expanses of canvas. Also, not many of the migrants had cameras.

There were other embellishments painted on wagon covers too. "Ho for the Pike Mines" was one such popular declaration. A few artistic migrants embellished their wagon covers with the imagined likeness of Pikes Peak. One was seen that displayed the massive profile of an elephant. The expression "to see the elephant" was popular in 19th century America. Loosely translated, it meant going west to see the sights, or to embark upon a new life beyond the Missouri River frontier. Among literate "go backs," the word "Busted," the expression "Busted by Thunder" or "Humbugged" replaced the optimistic and popular "Pikes Peak or Bust" slogans so flagrantly displayed on many westbound wagons.

A few inexperienced "go backs" fabricated log rafts on which they planned to float down the Platte River and back home to civilization. Some actually made it, despite the well-known shallowness of the Platte. Others got hung up on the river's abundant sandbars. A few were drowned. Those impoverished souls who walked back met unspeakable hardships. For those who died on the way, hastily excavated graves, always shallow, were prepared along the trails. One curious rationalization for "giving it up" after a mining failure was heard from a man who insisted that "the mines will not pay as well as staying at home with one's wife."

A second wave of gold greedy stampeders began to arrive in the spring of 1860. Family groups became more common in this second migration. Estimates suggest that between sixty and seventy thousand people went to the Rockies during the summer of 1860. Another estimate places some eleven thousand wagons on the roads to Denver during May and June of that year.

By no means all of those who found no gold turned bitterly back toward the East. Quite a few decided that they liked this strange new country, with or without easy riches. Because most nineteenth century Americans subsisted by farming, it seemed logical to file on agricultural homesteads along the Platte, Arkansas, and other rivers. Others started businesses that catered to both miners' and farmers' needs. A lesser number started towns of their own. And of course, there were those who went to work for wages, laboring as miners, clerks, handymen, bartenders, or teamsters.

Literally thousands of persons braved real personal privation as well as physical difficulty on the trails. At times there were serious dangers from the hostile environment. Prairie winters, including blizzards, the long waterless heat of summer, quicksand, swollen river crossings, disease, and accidental shootings took a greater toll of human life than the remarkably calm plains Indian tribes. The reason for this was that the Indians had little or no interest in gold. They thought that the gold seekers were not interested in the buffalo, and that both groups could co-exist. But sometimes frustrated Indians burned off precious prairie grasses and gold seekers' oxen went hungry.

Although tales of Indian depredations terrified the Pikes Peakers, few of them were discouraged by the far more serious prospect of death on the trails. In general their minds were closed. These things might afflict some other family, but never themselves. Inevitably, some ill prepared migrants simply got lost.

Since that first winter of 1858–59 had been an unsasonably mild one in eastern Colorado, the migration continued well beyond the period regarded as being safe for prairie travel. An increasing number of blue or white topped wagons lumbered over the trails until early December that year. Jerome Smiley said that "Christmas day was like June." An equally mild spring had allowed wagon traffic to be on the trails by mid-April 1859. Some migrants took their time, camping beside the streams while waiting for the grass to grow green along the route ahead. Opportunistic river town newspapers had printed little about the lack of either fuel or water on the western prairies.

One young woman, a pedestrian by choice, refused all offers of a ride and a few proposals of marriage from a number of lonely males on the trail. Two other ladies were seen driving their own backboard, pulled by two horses, all the way to Cherry Creek. They arrived safely, having experienced no discouraging incidents along the trail.

Probably the most original transportation innovations seen on the prairies were the wind wagons. At least three are known to have attempted the crossing. Irving Howbert described the one that passed their wagon on the Platte River Trail while crossing Nebraska. He said that its body was a common light duty spring wagon. Two men were seated in it. A single canvas sail billowed out from a mainmast above their heads. Howbert said that it was making "great speed" when it passed them. But anything that moved at a rate greater whan the usual ox pace of ten to fifteen miles a day would have seemed fast by comparison. Howbert also noted that when the Platte Valley winds had subsided a few days later, their slow moving ox teams passed up the dry land sailors.

One not too successful version of the wind wagon motif was advertised in the *Missouri Republican,* a paper that enjoyed a wide circulation in the river front embarkation towns. It stated that a Mr. Thomas of Westport Landing had constructed a twenty-four passenger wind wagon that could travel one hundred miles a day. Allegedly, the craft was wrecked by a prairie zephyr not far west of its starting point. The accident occurred on its very first voyage. The fate of any passengers remains unknown.

A third wind wagon left Westport Landing, Kansas, on April 19, 1859. Its fabricator had been subjected to considerable ridicule from the locals, and each evening a group would congregate at the waterfront "to see the idiot." The vehicle, apparently a modified buckboard with a single sail, was wheeled to the edge of town. At two o'clock in the afternoon, the winds met the operator's approval. He hoisted a sea bag into the body, mounted the seat and unfurled his canvas. Much to the surprise of the local Hirams, it began to move.

Colorado folklore insists that the contrivance made the crossing in six days — well almost. Its wreckage was observed beside the trail about twenty miles east of Denver. There were no blood stains and no body. Probably an overly-stiff wind gust flipped the brig of the prairie and the driver jumped clear to walk into town with the rest of the nameless throng. In April 1860, a windjammer from Nebraska made the crossing and actually arrived in Denver. One wind wagon was described as having wheels twenty feet high. A diary of the time told how horses became unmanageable when a wind wagon passed. Pet dogs, naturally, jumped out of their owners wagons, noisly chasing the strange conveyance until called back.

One of the best-documented accounts of a wind wagon that crossed to Denver appeared in the *Denver Times* for January 25, 1900. It stated that the conveyance was built in Oskaloosa, Kansas, by Samuel Peppard, who lacked the price of a conventional wagon and team. So he built a wagon body shaped like a skiff. It was eight feet long, four feet deep, and three feet

*Collection of George Foott*
This fine representation of a wind wagon was researched and drawn by Colorado artist George Foott.

across. The axles were six feet apart and all four wheels were of the same diameter. A ten-foot mast was fastened to the front axle and came up through the floor planking. Its rigging could accommodate two sails. In a high wind, only the smaller one would be hoisted. Both could be raised when prairie breezes were slight. It weighed a mere three hundred fifty pounds and could carry a crew of four with five hundred pounds of provisions. Doubting neighbors looked askance and with rural wisdom, dubbed it "Peppard's Folly."

With three companions, Peppard covered fifty miles the first day. Crossing Nebraska, they followed the Platte River most of the way before turning south toward Denver. On their best day, they covered fifty miles in a mere three hours, passing six hundred twenty-five teams in the process. Another time, the winds permitted them to travel ninety miles, and they considered that a "good day's travel."

Once a band of Sioux, presumably intoxicated, sighted them and a race ensued. When things became desperate, Peppard hoisted the larger sail, leaving the frustrated pursuers behind in a cloud of dust. One can only speculate at the treatment these braves must have received when they recounted their adventures back at the home village.

"Peppard's Folly" nearly made it to Denver. All went well until they had reached a point some fifty miles northeast of the Cherry Creek settlements. There, a whirlwind was sighted. When they attempted to lower the sail, the rope caught in the pulley and was severed, leaving no way to get the canvas down. A gust lifted the frail craft about twenty feet into the air before dropping it onto the rear wheels. Both wheels were smashed, but none of the men was hurt. Using the front wheels and some boards from the body, a rude cart was fabricated to move their supplies into Denver.

The *Missouri Republican* commented on the gold rush in their issue of March 21, 1859:

> Here they come by every steamboat, hundreds of them, hundreds after hundreds from every place — Hoosiers, Suckers, Corn Crackers, Buckeyes, Red Horses, Arabs, and Egyptians. Some have

ox wagons, some have mules, but the greatest number are on foot, with their knapsacks and old-fashioned rifles and shotguns; some with their long-tailed blues, others in jeans with bob-tailed jockeys; in their roundabouts, slouch hats, caps, and sacks. There are a few handcarts in the crowd. They form themselves into companies of ten, twenty, and as high as forty-five men have marched out, two-and-two, with a captain and a clerk, eight men to a handcart, divided into four reliefs, two at a time pulling the cart. . . . onward they move, in solemn order, day after day, old and young, fat and slender, short and tall, handsome and ugly, the strong and the weak.

Irving Howbert was among the most observant of the migrants. He noted that there were few families in the rush. He also noted that most of the "Peakers" were single men or heads of families whose families would follow them west at a later time when homes had been established for them on the raw, new frontier. But occasional wagons were seen with women inside. Most of these family vehicles were recognizable at a distance by the milk cows that were driven or led behind the tail gate. Often such vehicles had a crate or two of poultry lashed onto the sides.

Because the early migration was made up primarily of males, each Argonaut had to do his own cooking or go without. Everything was either boiled, fried, or consumed raw. The Pikes Peakers subsisted on stomach-abusing diets of coarse pancakes, bacon, beans, and coffee. Canned foods, when available, provided the only variation from an otherwise monotonous culinary experience. Dutch ovens, coffee pots, frying pans, and tin plates became recognized essentials of the miner's duffel. Thousands of tin cans were strewn along the right-of-way, a condition not unknown to contemporary roadways.

Most cooking was done over hot but brief open fires, fueled with dried buffalo chips. One wagon was seen with a small sheet iron stove lashed to the tailgate. Three times each day, the little stove was set up, fueled, and the appropriate meal prepared. Wood for fuel was relatively scare. The abundance of willow and cottonwood trees that now line the Platte Valley did not exist at that time. Groves of trees were miles apart.

Reaching a stand of deciduous trees was an exciting event in the lives of the migrants.

The cameraderie of the trail was sometimes the source of strong and enduring bonds of friendship. People who had traveled and endured hardships together on the prairie, stayed together, forming lasting partnerships in the mountains.

Those Kansas frontier towns at the eastern end of the several

*Collection of Frances and Richard Ronzio*

Famed landscape artist Albert Bierstadt was one of the firm of Bierstadt Brothers of New Bedford, Mass., when they photographed these Pikes Peak emigrants at St. Joseph, Missouri, in 1863.

routes to the mountains were small and often very new communities. All had been hard put to survive the Panic of 1857. A prospective flood of gold seekers meant prosperity for the town that could attract them. Each community exaggerated its own advantages as starting points for the routes to the mines. The "Peakers" were charged outrageous prices for a varied assortment of mining aids, most of which were useless. One, called a goldometer, was supposed to reveal where one might dig for that elusive element.

Some of the travelers carried maps that dated back to the 1820 expedition of Stephen H. Long — maps that labeled eastern Colorado as the "Great American Desert." Tragically, others purchased guide books written by people who had never visited the Great Plains. Westbound migrants were badly deceived by these irresponsible writers, who, in turn had been retained by unscrupulous Missouri River town merchants to assure that the Argonauts would be funneled through their communities.

There were a few guide books in print as early as December 1858. In retrospect it would appear that each succeeding author went out of his way to embellish the unlikely accounts of his predecessors. One of them, called *A Handbook to the Gold Fields of Nebraska and Kansas,* was co-authored by William Newton Byers and John H. Kellom, who at the time had never been to the Pikes Peak Country. It sold for fifty cents. In all, twenty-eight Pikes Peak guide books had been published by 1861. Nearly all of them listed the supplies one should take with him. Most lists included a skillet, tin plates, Dutch ovens, coffeepots, a cook stove, butcher knives, a tent, and sixteen pairs of blankets. A wagon cost $75, and an adequate team to pull it could be had for $125.

One guide book, written by a hack who obviously had never been west of the Hudson River, asserted that gold could be had at Pikes Peak by climbing to its 14,110-foot summit. There you were instructed to cut down about fifteen stout trees to be lashed together with rawhide, forming a flexible raft. (No trees grow above the 10,800 foot level, and no gold was found near Pikes Peak until 1891.) Anyway, gravity would allow the gold seeker's raft to skim down the steep slope, its corduroy-like under surface peeling off layers of gold as the decline progressed. Several thousand dollars worth of the yellow metal could be scraped off the under surface by the time one had descended to the level of the plains.

One other closely-related tale concerned a group of prospectors who had dragged a log barge to the summit of Pikes Peak. Horizontal strips of iron were attached to its underside, and the hull was loaded with rocks. When the conveyance was pushed off the top, the metal bars stripped the gold from the extruding mountainside. After the clumsy device had made its way down, a ton of gold shavings was retrieved from beside the path.

Another publication insisted that gold nuggets "lay glittering around Colorado's high peaks like a necklace." Little wonder that Colorado's gold rushers cursed the guidebook writers. Consequently, few authentic Pikes Peak guidebooks have survived. Most people quickly consigned them to camp fires. Others used their worthless pages for tinder to start cooking fires. Nearly all of the remaining samples are in museums. The general lack of adequate information about trail conditions was a genuine tragedy. Some people left Westport with provisions for less than a week, woefully inadequate preparations for the six hundred miles to be traveled at ten-to-fifteen miles a day.

In all fairness, it should be noted that a few of the guidebooks were good. John C. Fremont's guide to the Oregon Trail, actually written by his wife, Jessie, was one such exception. Although published many years before, it was accurate where trail conditions and distances were concerned. Another book that traveled west in gold seekers' wagons was the Bible. An agent of the American Bible Society was active at Leavenworth, Kansas, as early as 1859, offering free copies of the Pocket Testament to those who would accept them. The National Board sent one thousand copies of the Testament and one thousand Bibles to be distributed by "the few earnest

Christians who mourn for the desolation of Zion." The number of these texts that actually reached Cherry Creek cannot be determined.

Missouri bordertown businessmen on the trails east of Colorado became rich selling guidebooks and outfitting the throngs of lighthearted gold hunters. Merchants beamed with anticipation at the streams of greenhorns that moved through their shops. Whether gold existed or not, they welcomed the business as each new rumor sent more blanket-laden optimists to posture in their stores. River town sharpers sold worn out, footsore oxen and spavined horses to the migrants. Sometimes they represented donkeys as mules to gullible Pikes Peakers, too anxious to get out on the trails to realize that they had been cheated. One sizeable party was so incensed that they returned to Plattsmouth, Nebraska, threatening to burn the town's business section to the ground. Local merchants, they insisted, had fabricated the gold excitement story in order to create a demand for their goods.

Hard liquor was easily one of the most popular commodities vended by Missouri Valley outfitters. It was widely believed that whiskey was an effective antidote for snakebites. And so, it was said, nearly all westbound wagons carried in their cargoes a case of whiskey, and a box of snakes. Beginning in 1859. handsomely embossed flasks of the fiery liquid, became available. With some minor variations, the embossing showed a gold seeker on the front side and a man shooting a stag on the rear surface of the bottle. Below the flask's shoulder, the popular logo "For Pikes Peak" appeared in raised letters.

In capacity, the flasks varied from a half pint to a quart. Their colors were equally different and included dark brown, blue, green, amber and clear. Several of the earlier bottles had open pontils. So far as is known, the last of these unique bottles was made in 1872. All of the surviving Pikes Peaks flasks have become valuable and are highly prized by collectors of antique glassware.

All of the trails used the Missouri River frontier as a jumping off place for the trip to the "Pike Mines." Those frontier towns that were the principal embarkation points were located along the river front and could be reached easily by stagecoach, railroad, or steamboat. Each sternwheeler that came up the river was loaded with "Peakers." Singly, by tens, and by twenties they came. But not all who came were out-of-work farmers. Some were former traders or trappers. Others were soldiers of our Indian-fighting army, seeking discharges in the field from western military outposts. An impressive number were merchants with wagons loaded with goods for sale in stores they would establish on the new frontier. One, a Nebraska dairy farmer, deplored the reported lack of butter in the gold rush towns. Shortly after Christmas, he loaded a wagon with butter and started for Denver. An unexpected January thaw shattered his dreams. An enormous grease spot on the prairie was visible for weeks.

At the major rivers, long caravans of wagons waited impatiently for log ferries to take them across. Although the migration started in the spring, most people waited for the mud

*Collection of Karen and John Eatwell*
These rare Pikes Peak commemorative flasks of whiskey were sold by Missouri Valley merchants for about two decades after 1859.

to dry up in Iowa. A few wallowed through it. If they chose the Smoky Hill route, they could be at Cherry Creek by mid-June, if they started early enough and if their luck held out.

As a substitute for other accommodations, many brought tents that could shelter them while they camped beside the icy streams. Others threw together crude brush lean-tos that were abandoned when they moved on. Most of the hardier souls slept inside or under their wagons, or merely rolled up in a blanket on the open plains, a process called "sleeping on prairie feathers."

From time to time, western army units patrolled in an effort to keep the trails safe. Known bands of Indian troublemakers were sometimes pursued. But the vastness of the country, the known ability of the natives to be inconspicuous when the occasion warranted it, plus the sparse numbers of troops, mostly inexperienced, that were available for frontier duty, doomed most such security efforts to failure.

Indian depredations, it was agreed, would eventually be punished by the army. Among the military leaders who led punitive expeditions were Colonel William S. Harney in 1855, and just prior to the Civil War, Col. Edwin V. Sumner in 1857–58, Lt. Col. Albert Sidney Johnston in 1858–60, Maj. John Sedgwick, who patrolled in eastern Colorado from the Santa Fe Trail north to the South Platte River during 1857–60, Col. John M. Chivington, who fought Indians here after 1861, and Lt. Col. George Armstrong Custer after 1865.

Those wagons that traveled in groups fared best. When confrontations did occurr, distribution of gifts among the tribesmen was sometimes effective. On those occasions when marauding Indians were successful with their attacks on small groups of migrants, the story was quickly spread by survivors and, understandably, embellished with each telling. The vicious, painful, and at times erotic nature of the acts visited by Indians upon their unfortunate captives provided no solace. The fact that captives from enemy tribes were treated in a like manner was never considered by those terrified families being fed a steady diet of atrocity stories in frontier supply towns. Cruel atrocity stories aside, there were many families that made the journey across to Pikes Peak without ever seeing an Indian. Diseases such as dysentery and cholera (the big cramps) took a far greater toll than Cheyenne or Arapaho warriors.

# 6.

## GOLD RUSH DENVER AND ITS
## CONTEMPORARY RIVALS

DENVER, as it soon became known, had been started to meet
the demands of the gold rush, and to provide needed services
and desirable luxuries to those who sought quick riches within
the several mountainous mining areas to the west. Wherever
Americans are gathered together on a frontier, someone will
inevitably call a public meeting to organize a town. Following
that, at least half of the people will declare themselves available
to run for Congress.

Inevitably, other parties had followed the original three
groups until there were between six hundred and eight hundred
persons on the scene by the onset of the winter of 1858–1859.
Most, but not all, had settled near the point where Cherry
Creek flows into the South Platte River. In the earliest days,
the common name for this spot was "The Confluence."

By the time the first snow came in late December 1858,
there were many tents, several cabins were already nearing
completion, and a few others were soon finished and occupied.
Several pre-existing cabins stood near the Platte along a
thoroughfare known as Indian Row. In the center of the dry
bed of Cherry Creek stood the cabin of Jesus Abrieu, the
son-in-law of Carlos Beaubein.

Only five women were noted as being in Denver in 1858.
One was John Smith's "squaw," Wa-po-la. The second was
Mrs. D. M. Rooker, who had come over from Utah. Uncle
Dick Wooten's wife had arrived from New Mexico on Christmas
day, 1848. Smiley's *History of Denver* mentions a Mrs. Smoke,
whose husband operated Denver's first "hotel." Of the original
five only Katrina Murat, wife of Denver's first barber, chose to

remain. But by the end of that year about thirty-five additional ladies had arrived. Several, including Augusta Tabor and Mrs. William N. Byers, arrived the following year. Another of the earliest ladies was Mrs. Joseph Wolff, wife of a newspaperman who had been expelled from Wheeling, Virginia, (there was no West Virginia until 1863) for publishing an anti-slavery paper. Mrs. Wolff, aghast upon seeing the primitive accommodations in which she was expected to live, refused to cook. Allegedly, the lady also withheld other wifely favors until better facilities were provided. Mr. Wolff lost little time in getting a fine roof on their cabin, and hastily improvised other improvement until domestic tranquility was restored.

Deterrents to normal family life remained throughout the first decade. No one knew precisely how long the mineral deposits would last, and many husbands consequently did not send home for their wives. The trip west was long, costly, and sometimes marred by disease and violence.

In an earlier chapter, the historical priority of Montana City was noted along with the decision of John Easter's party to abandon it. Subsequently, both the people and some of the structures were removed to a site ten miles north. There they laid out a second town. It was called St. Charles. "Lying John" Smith and William McGaa, "squaw men" and Indian traders who lived in tepees close to the mouth of Cherry Creek, were invited to participate in the St. Charles real estate venture. Between them, Smith and McGaa had considerable influence among the Cheyenne and Arapaho people.

McGaa was the well-educated son of an English nobleman who ran away from home, set out for America, and made his way to the far west. Smith, a native of St. Louis, left his home in 1826 to work for the American Fur Company. He attached himself to the Cheyenne nation, becoming a chief of great authority. Later he abandoned Indian ways to become a trapper, trader, and guide. He was noted for his remarkable resemblance to Andrew Johnson. John Smith was one of those persons for whom telling the truth held little fascination. Because his word meant so little, the early settlers dubbed him "Lying

An early artist took artistic license with this view of pioneer Denver. Cherry Creek comes in from the lower left, while the South Platte River runs across the painting horizontally.

Here is the much-changed point where Cherry Creek still joins the South Platte.

John." At a later time Smith was run out of Denver for breaking his wife's back with a three-legged stool.

And so, St. Charles became the second town in priority of time to spawn within present Denver. The St. Charles Association dates from September 24, 1858. The townsite was bounded on the northwest by the South Platte River, and on the southwest by Cherry Creek. It occupied a substantial part of contemporary downtown Denver. But St. Charles remained nearly empty until the arrival of Gen. William Larimer in November 1858. Only one cabin was ever built at St. Charles.

Larimer was a native of Pittsburgh, Pennsylvania. He had been a banker, a real estate promoter, a general in the Kansas militia, and president of a railroad. In common with many others, he had suffered severe financial reverses after the Panic of 1857. To attempt a reversal of their ill fortunes, Larimer and a small group of friends formed a prospecting party at Leavenworth, Kansas. Larimer's associates met with him for the first time in mid-November of 1857. Soon they were joined by persons from the nearby towns of Lecompton and Oskaloosa.

Prominent among the Lecompton contingent was Edward Wynkoop, later a major and then a colonel in the Colorado Volunteers. Although Wynkoop headed the Lecompton party, Larimer himself was the accepted leader of the total enterprise.

Nearly a hundred person had expressed an interest in going, but when the day of departure arrived, only six people showed up to begin the trek westward. Along the way five others joined up. This latter group consisted of people who had gone to Kansas Gov. James W. Denver, requesting that he appoint them to be the first officials of Arapahoe County,. Among the entourage were a judge, a territorial commissioner and a Kansas sheriff. It now seems probable that the officials were included to dignify what Larimer was about to do. Among his other talents, Larimer was an altogether capable town promoter. Forty-seven days were required for the party to cross from Leavenworth to Cherry Creek.

In the meantime, the weather turned chilly, causing the original St. Charles Town Company people to depart for the milder winter climate of Kansas. They left behind either one man or a small group to sell lots and to guard their holdings. Contemporary accounts disagree on this point. But there is no disagreement concerning what transpired next. Gen. Larimer moved in on November 22, and literally jumped the claim of the St. Charles holdings. He or someone in his party plied the remaining resident(s) with whiskey and promises. There were threats, too. Badly outnumbered, the opposition capitulated, bowed to the inevitable, and Larimer took over.

It was Larimer who laid out Denver's first streets, but without regard to the points of the compass. Rather, they were planned in relation to the flow of Cherry Creek. Even today, Denver's downtown streets run at odd angles as compared with the remainder of the city.

To disguise and complicate the take-over, Larimer changed the St. Charles name to the Denver City Town Company, honoring the territorial governor of Kansas. Larimer assumed that Governor Denver would enjoy seeing his name on a map, thereby causing him to side with the Larimer party. Unknown to Larimer, Denver had already resigned and was on the way to Washington. But the town's name, once established, persisted.

James W. Denver was a native of Winchester, Virginia. He grew up on a farm near Wilmington, Ohio. He sometimes practiced law. At other times, he edited newspapers, including the *Platte Argus* at Plattsburg, Missouri. When our war with Mexico began in 1846, he was commissioned a captain. Denver served from Vera Cruz through to Mexico City.

Following the cessation of hostilities, Denver crossed the plains with the Forty-Niners to Sutter's Mill. In California he was elected to the state senate in 1851. When prospectors were caught in deep snows while attempting a winter crossing of the Sierras in 1852, Denver headed up the rescue mission. Somehow this led to differences with Edward Gilbert, editor of the *Alta Californian,* who challenged Denver to a duel. Gilbert dressed in green to appear less conspicous when the two met at Oak Grove near Sacramento. Their weapons were rifles at forty paces. Denver, an accomplished marksman, deliberately missed. But Gilbert came close and loudly demanded a second shot. Gilbert missed again, Denver didn't. Editor Gilbert died where he fell with a bullet in his chest.

After a time, the duel was forgotten and Denver was elected California's secretary of state in 1853. The following year, he was sent to Congress where he quickly allied himself with forces pushing for the transcontinental railroad. President Buchanan appointed him a commissioner of Indian Affairs in 1857. The following year, he was appointed to be the governor of Kansas Territory. It was during his tenure in that office that the Pikes Peak gold excitement erupted in the western part of Kansas' Arapahoe County.

The Buchanan administration had many troubles, culminating in one of those rare instances of a sitting president being denied renomination by his own party. President Buchanan, our only bachelor chief executive, faced his many problems by alternating between crying jags and shrill prayer, two characteristics that have delighted his more critical biographers. On virtually his last day in office, Buchanan signed the bill creating Colorado Territory.

Abraham Lincoln, Buchanan's successor, appointed James Denver a brigadier general and assigned him to serve under General Sherman in the Civil War. Then in 1876 and again in 1884, the Democrats advanced Denver's name for the presidential nomination. The efforts failed when his opponents dredged up the duel and used it to deprive him of the nomination. When Denver visited his namesake city in 1875 and again in 1882, his personal letters complained of the lack of public attention accorded him. He died in Washington, D.C., and was buried at Wilmington in 1892.

Curiously, General Larimer chose not to spend his declining days in Colorado. For many years he coveted the governorship of Colorado. Later, he tried to be elected mayor of Denver. When both offices eluded him, he left the mountains for home. Larimer died in 1875 at home on his farm at Leavenworth, Kansas.

Back in western Kansas, the two original towns were followed by a third emerging hamlet. It was called Auraria and was named both for Auraria, Georgia, and for the fact that Auraria is the Latin word for gold. Such a choice was largely a nostalgic selection by the homesick Russell brothers from Auraria, Georgia. More precisely, the Auraria name came from a suggestion by Dr. Levi Russell.

On October 20, 1858, five weeks after the beginning of St. Charles, a meeting was held to complain about the inflated cost of homesites across Cherry Creek. Among those who gathered were the Russell brothers, William McGaa, John Smith, and a number of equally-concerned parties. Their meeting was convened on the southwest side of Cherry Creek. Soon they had decided to start another new town. Quite an impressive number of people chose to follow the protesters across the creek to Auraria, mostly because the price of building lots in Larimer's Denver continued to be very expensive. Probably as a slap at St. Charles, town lots in Auraria were free to anyone willing to locate and build.

In November 1858, as winter approached, Green and Levi Russell left Auraria to spend the cold months basking in the warmer Georgia climate. Oliver, the remaining brother, stayed behind and wintered at the new town. Brother Oliver was a musician of sorts — a fiddler who found himself in considerable demand at all kinds of social occasions on both sides of the creek.

By the time the new year rolled around in 1859, an alert observer counted fifty cabins in Auraria. But, he noted, there were only half that number in Larimer's Denver City. Auraria opened its own post office on January 18, 1859. It was discontinued on February 18, 1860, when Auraria merged with the rival Denver City.

At midnight on April 5, 1860, a bright moonlight night, the rivalry between the two Cherry Creek towns was resolved by a ceremony at the center of the Larimer Street bridge. Appropriate speeches were heard between toasts, and firearms were discharged into the clear night air. Finally, the moment of truth arrived as Auraria and Denver City were merged into a single community. Economic rivalry from Golden City had pushed the two competing villages into a hasty union of convenience and expediency.

At the time of their merger, the combined population at Denver and Auraria has been estimated at two thousand. Denver's people provided a variety of services to an additional 34,277 who lived in the nearby mountain camps. Denver-1860 was composed of shacks, frame buildings of the most fragile variety, and an assortment of tents. Except for the tents, most of the business structures sported fashionable false fronts. Streets were neither lighted nor paved, and only Blake and Wazee streets gave even the appearance of being public thoroughfares.

Among the earliest arrivals at Cherry Creek were two men named Charles H. Blake and Andrew J. Williams. They had come across from Iowa with four ox-drawn wagons filled with trade goods. They reached Cherry Creek in October 1858, unloaded their general merchandise, and opened the first business venture in Auraria. Three months later, they moved across Cherry Creek to St. Charles and set up a thirty-by-one hundred-foot structure of cottonwood logs. It stood along the north side of present Blake Street near Fifteenth. Their actual store occupied only a small portion of the structure. The remaining space became a crude hotel, known at first as the Blake and Williams Hall. It had no permanent walls to separate the sleeping cubicles. Only sheets of canvas or burlap nailed on flimsy wood frames separated these rudest of accommodations. Neither material could screen off the raucous noises of nearby sleepers. Snoring, coughing, belching, and other nocturnal sounds were clearly audible to any of the unfortunates who attempted to sleep there.

Across the room, a three piece orchestra played for nightly dancing, adding to the cacophony. At the front for accessibility, a saloon and twelve-table gambling den functioned noisily around the clock. Each evening when all its noctural functions were in full operation, the noise could be heard half a mile away. One sensitive pioneer commented that the Blake and Williams place sounded like the arrival in town of an entire

This view of early Golden shows Lookout Mountain at the far right.

Here is a modern view of the same scene.

Italian grand opera troupe. Meals were served in a space next to the saloon. Because the building lacked a proper foundation, the floor was earthen.

Due to the vastness of the structure, occupying fully a quarter of the block, it soon adopted the name of the "Elephant Corral." The unusual name may have been suggested by the Elephant Corral at Council Bluffs, Iowa, or by the popular expression, "I've seen the elephant," meaning that one had traveled to the West and seen its many wondrous sights.

Rates charged were said to be no higher than those assessed by other "first class hotels." With considerable tongue-in-cheek, Horace Greeley once referred to the Elephant Corral as the "Astor House of the gold regions." The Elephant Corral went up in smoke during the great Denver fire of 1863.

In lower downtown, a log ferry was built to facilitate crossing the South Platte River. Predictably, the street that led down to it was called Ferry Street. Later its name was changed to the less-interesting Eleventh Street. Just off Ferry Street in lower Auraria stood a line of log cabins known as Indian Row, built well before 1858. Next to them was a newer but still early landmark, Denver's first brothel. Ada Lamont had ventured west with the first wave of migrants, not to mine gold, but to mine the miners. In a society where women were either a rarity or a bargain, Ada was quite naturally described as being "beautiful." Originally her intentions had been altruistic when she accompanied her clergyman husband to the gold fields to harvest the souls of Godless miners.

But somewhere along the way, the Reverend Lamont brought about the "salvation" of a migrant woman who was said to have been easy on the eyes and of easier virtue. The two of them departed together, probably to seek spiritual solace elsewhere. Ada searched diligently, but never found them. So for the next decade she supported herself on Indian Row, selling the services of a bevy of courtesans, ministering to the glandular needs of lonely miners, Ada's Place was Denver's first love store, a sign that civilization had arrived.

"Count" Henry Murat was Denver's first barber. During

*Collection of Freda and Francis Rizzari*

Cheney's Saloon and Billiard Hall stood on Washington Avenue in Golden. The
building of W. A. H. Loveland, transportation pioneer, is at right.

*From the Collection of Evelyn and Robert L. Brown*

The bluffs of Table Mountain still dominate Golden's skyline. Inset shows saloon
tokens.

the visit of Horace Greeley, the famed editor entered Murat's shop for a shave. A truly thrifty man, Greeley felt he had been shaved in more ways than one. He was horrified by Murat's five dollar price. Murat built a log hotel over which the area's first American flag was flown. The flag was the handiwork of "Countess" Katrina Murat. The Murats claimed kinship to the emperor Napoleon.

Inevitably, a network of towns sprang up to house the would-be miners. Four such towns were within or close to the contemporary limits of Denver. Arapahoe City was situated to the south of West Forty-Fourth Avenue, east of present McIntyre Street. Briefly, it was home to George A. Jackson, and was the place from which he began his prospecting trip into present Clear Creek County. When his sojourn had been completed, he returned to Arapahoe City. Originally, Tom Golden had lived there, too. Arapahoe City was platted by George Allen, whose house still stands at the site.

West of Arapahoe City, the new town of Golden had replaced the earlier Golden Gate City. Although some sources credit W. A. H. Loveland with being Golden's founder, the real instigator was George West, who worked with the Boston Company. Together, they named the community for Tom Golden.

On December 14, 1858, General Larimer crossed the Platte and laid out a town called Highlands. It was located on Denver's north side. It grew up in the vicinity of Twenty-Sixth Street and Federal Boulevard. Its city hall stood on the corner where the fire station is currently located. It was an agricultural town, where farmers raised produce for sale to prospectors on their way to the mountains. It was merged with Denver in August 1860.

Coraville, little known, short lived, and obscure, was the last of the Cherry Creek settlements. It was situated on the east side of Cherry Creek at Fourteenth and Larimer streets. Although tiny, Coraville had its own post office, located in the express office of Jones and Russell's Leavenworth and Pikes Peak Express Company. Mathias Snyder of Virginia was the postmaster. Cora Snyder, his wife, was the source of the Coraville name.

*Western History Department, Denver Public Library*

Gen. William Larimer sat for this portrait in 1860.

Gen. James W. Denver, governor of Kansas Territory at the time of the gold rush.

Actually, the place was little more than a postal facility and not a real town in the usual sense. Coraville's post office was established on March 22, 1859, and was discontinued on June 25, 1859. Across the creek, Auraria's post office began to function on January 18, 1859 and was discontinued on February 11, 1860.

And finally, perhaps to prove that it had earned its place in the sun, when Christmas 1860 rolled around, Denverites celebrated the occasion with a public wrestling match. That same day, Uncle Dick Wootton returned from a trip to New Mexico with a wagonload of Taos Lighting. Although the precious cargo was intended as stock for the soon-to-be-opened Wootton Western Saloon, glasses of the powerful brew were freely distributed to the Yuletide throng. Truly, civilization had arrived.

# 7.

## APPROACHES TO THE GOLD FIELDS

THE WEST'S FIRST TRAILS were made by the buffalo, deer, and antelope of the plains and mountains during their seasonal food seeking migrations. The western Indians frequently followed these crude paths, particularly the Utes, a Shoshonian people who lived in the mountains and on Colorado's western slope.

Some of the plains' trails north from Santa Fe had been used by Spanish conquistadores in the 1500s. Official American explorers, traders, and trappers traversed these same well-known paths across the prairies in the 1800s and opened others within Colorado's Rockies. A few of the better routes survived the transition to stagecoach travel. Other paths, not always the best choices, were widened, cribbed, and bridged to accommodate the hordes of gold rushers who poured into Colorado in 1858–1859. Some people came up to Cherry Creek by way of the old trapper's trail from Taos. There was also an early trapper's route from the Arkansas River to Fort Laramie. Because it crossed the Platte River, access to the Cherry Creek settlements was easy from that direction.

Undoubtedly the best-known among all of the roads used by people on the way to the gold fields were the already-established Santa Fe and Oregon Trails. For the first forty miles west of Independence, both trails used the same route. West of present Topeka, they split: the Santa Fe road went southwest toward the Arkansas River, while the Oregon Trail moved northwest into Nebraska, following the North Platte River toward South Pass.

All three of the earliest parties followed the Arkansas River

W. H. Jackson painted this scene of wagons on the Oregon Trail.

The Oregon Trail's prominent landmarks are evident near Scotts Bluff.

route, the northern branch of the Santa Fe Trail, nearly to the mountains. Their prospecting frequently began there. More specifically, the Argonauts followed the Santa Fe road only as far as the ruins of Bent's fort. There they turned north on the Old Cherokee Trail toward Fountain Creek.

For those who chose the Oregon Trail, about six weeks were required to travel from the Missouri River to Julesburg in the South Platte Valley. Wagons usually traversed the south side, although the Mormon's 1847 wagon ruts were still visible along the Platte's northern shore. Here and there, the trail ran close to the river, while at other times it might swerve away by as much as a mile. At Fort Kearney, the feeder wagon roads from St. Joseph and Council Bluffs joined the main trail. The presence of several forts, farms, and ranches along its right-of-way made the Oregon Trail a desirable choice. West of the Missouri River, Fort Kearney was the most used secondary assembly point. Understandably, each trail had its partisans who claimed the best road conditions, water, fuel supplies, and abundant prairie animals for food.

When Irving Howbert's family followed this road, he described an almost continuous procession of canvas-topped wagons, extending for miles in front and far down the valley behind them. Howbert's estmate was that if all of these wagons could have been put together, the line would have covered a fourth of the trail or about one hundred miles of the route between Fort Kearney and Denver. In addition to wagon traffic, some people came pushing wheelbarrows, go-carts, and Mormon-type handcarts. Others walked through the snow and mud of early spring or the dust of summer.

Quite apart from the myths perpetuated by bad fiction and worse television, the Plains Indians were rarely a problem to the earliest gold hunters. Supposedly, America's total Indian population was about five hundred thousand in 1492, a number that has remained almost unchanged up through the most recent census figures. For the sake of comparison, this is the approximate population of contemporary Denver. The Indians of the Great Plains constituted a much more modest number.

When such a number of people are dispersed between the Missouri River and the Pacific Coast, we should not wonder at the fact that many westbound travelers crossed the frontier without ever seeing an Indian. Only in the imagination of western hack writers did thousands of painted savages attack, scalp, and eviscerate the Argonauts of 1859. At first the Indians thought that we were not interested in the buffalo, and that we could all live together in harmony. The real prairie killers were cholera, scurvy, and diarrhea.

Oxen were the preferred beasts of burden chosen to pull westbound wagons across Kansas, Nebraska, and eastern Colorado. They were stronger than horses and much better suited to the dietetic rigors of prairie travel, where feed other than grass was unobtainable west of the Missouri River settlements. Oxen could move a wagon at a steady pace of about fifteen miles a day. Moreover, oxen were far less attractive to the Indian than horses or mules. For eating, buffalo were available with less hassle, and their flavor was not unlike that of the ox.

Where river crossings were difficult or dangerous, log rafts masquerading as ferries took wagons and teams across swollen portages at exorbitant rates. Here and there, an enterprising soul might set up an illegal shack and charge a dollar a wagon as toll for using the road. Mass indignation, open threats, and outright refusals to pay discouraged the practice among all but the most tenacious. Other opportunists got away with charging for the use of waterholes. Where trees were scarce, wood wagons appeared to vend that commodity for emigrant cooking fires. There were few cottonwoods, and they were soon cut. As a last resort, one could usually pick up enough dried buffalo chips to fuel a fire. Incidentally, the scars of these much-used campsites are still visible today at approximately fifteen-mile distances along the Oregon Trail.

Inevitably, one of America's most cherished institutions, the saloon, also went west. Most trailside saloons consisted of little more than an assortment of intoxicants displayed on the tailgate of a wagon parked beside the road at carefully chosen locations. Hilltops were preferred, since thirst was induced among those whose job it was to push and haul beside the oxen on the steeper grades. And as the crest of a rise was reached, there stood the oasis waiting to minister to the parched throats of exhausted, sweating Argonauts.

One other route to Denver was shorter than either the Santa Fe or Oregon trails. It was called the Smoky Hill Trail. The Smoky Hill route is a very old one, dating from 1855 when a surveying party crossed the Great Plains and laid out a wagon road from Kansas to Bent's fort. Since it followed the Kansas and Smoky Hill rivers, it came to be known as the Smoky Hill Trail. A number of westbound migrants traversed it in 1856. Its greatest virtue was the more direct, almost straight line approach to the gold regions. Its disadvantages were lack of water and the presence of Indians. The Smoky Hill road followed the Kansas River through Junction City to the Smoky Hill Fork, a stream that rises in eastern Colorado. Zebulon Pike, Colorado's first official explorer, was responsible for the Smoky Hill name. The occasion was his 1806–1807 expedition.

Beginning at either Atchison or Leavenworth, Kansas, the Smoky Hill Trail split off from the main route at least three times, affording a choice of roads to Cherry Creek. Called the north, middle and south routes, they converged again at a point about twelve miles south east of Denver. Here they merged and entered the Cherry Creek drainage together.

In general, the western end of the trail traversed the route of the present Parker road into Denver, following the sandy bed of Cherry Creek. It entered the city by way of contemporary Cherry Creek drive and Speer Boulevard. East of Broadway, it turned north to the end of the trail Pioneer Monument at the intersection of Broadway with Colfax Avenue.

The so-called Republican River Road paralleled that stream as far as Big Sandy Creek. There, it crossed over to use the regular Smoky Hill route. Horace Greeley wrote of the "dreadful lack of wood and water" on the South Fork of the Republican River.

Along the South Fork of the Smoky Hill River in Kansas, a trail of Indian origin had followed the waterway before 1859.

The two swales on either side of the Spanish Bayonet plants are the scars of the Smoky Hill Trail.

On the Smoky Hill Trail, the original Four-Mile House has been restored by Denver's Parks Department.

It ran southwest to Bent's new fort, Big Timbers. Later, the trail was used by the Butterfield Overland Despatch stagecoaches, and by a branch of the Pony Express from Julesburg. Much of the Smoky Hill Trail was more than a mile wide, a condition dictated by the need to find adequate forage for oxen and mules, as well as campsites and water for people.

The Smoky Hill road traversed many miles of dry country, elevated and remote. Some of it was only semi-arid, while other parts crossed a virtual desert. In common with other pioneer routes, it followed the upland, mudless ridges during wet spring days, and the lower contoured valleys when conditions were dry. Its crooked wanderings might appear to the untrained eye to have been laid out by a contingent of dedicated inebriates.

The Middle Smoky Hill branch, the aptly named Starvation Trail, was easily the most desolate and dangerous of the three. It lacked water for most of the year. Yet because it was the oldest and was the most direct route, people followed it, encouraged by trail town merchants in Kansas. A greater number of people expired from hunger and thirst on the Middle Smoky Hill than from Indians.

The *Rocky Mountain News* once reported that two men had survived on the Smoky Hill road for nine days by eating prickly pear cactus and hawk meat. A reporter for the paper told of passing between ten and fifteen human bodies beside the trail. Two other men were reported to have traveled one hundred fifty miles with no other nourishment than melted snow water from two unexpected storms. The trail was lined with broken wagons, dead oxen and mules, much abandoned equipment, and a number of unmarked graves.

Probably the most pathetic story of the entire Pikes Peak Gold Rush details the experiences of Daniel, Charles, and Alexander Blue of Clyde, Whiteside County, Illinois. The Blue brothers started for Pikes Peak with little more than a burlap sack, a carpetbag, and high hopes. Two other men brought their party to five. Their hasty preparations were both superficial and inadequate for the rigors of trail life that lay ahead of them.

They left Illinois on February 22, 1859. Rail connections took them to St. Louis and a boat trip down the Missouri River placed them at their embarkation point, Westport Landing. Here the number in the party was swelled to fourteen. On March 6, they began walking west toward Lawrence, where they purchased a pack horse. In Topeka, two hundred pounds of flour was loaded on the tired animal's back. Six hundred miles of the treacherous Smoky Hill Trail lay ahead of them. Two others joined them at Topeka.

Four days later, they lost the pack horse. All supplies were now transferred to their own backs. For eight more days they stumbled along. Two discouraged men now left the party. Armed with only a shotgun, the remaining seven hunted for animals for ten exhausting days. At the end of that time three additional men departed, leaving only the three Blue brothers, their cousins, and a man from Chicago named Soley or Soleg.

On Bijou Creek, a tributary of the South Platte River, Soleg died of exhaustion. Since they had now been without food for eight days, the body was eaten. In a manner of speaking, Soleg "lasted" for eight more days. They were now about seventy-five miles east of Denver.

Despairing, Alexander Blue penned a farewell note to his wife on April 18, 1859. After suggesting that the others sustain life by eating his body, Alexander died. Two more agonizing days passed before Charles and Daniel could bring themselves to begin devouring their brother. Now much refreshed, they traveled about two miles for each of the next several days. Everything but the shotgun had now been discarded.

Several days passed before Charles informed his brother that he could not continue. He collapsed of exhaustion and malnutrition, but lived on for ten additional days. Following Charles' death, Daniel later recalled feeling both depressed and alone. He remained beside his brother's body for three days of mingled grief and hunger pangs before he began to feed upon the corpse. As he recalled the event at a later time, he felt as though his stomach had turned to stone.

At times he lost consciousness and assumed that he too

was about to die. Fortunately, three passing Arapaho braves found him and carried him to their village. A squaw bathed and fed him both raw and roast antelope. On May 4, the Indians removed him from their lodge and took him to the Twenty-Five-Mile House station of the Leavenworth and Pikes Peak Express Company. This location was probably the Parkhurst Station, five miles southeast of Parker.

During the three months since he had left Illinois, Daniel Blue had traveled about nine hundred miles, covering nearly half of the distance on foot. His journey was completed when he rode the stagecoach into Denver on May 12. Some of the deserters made it too, but only five of the seventeen that had started together actually completed the journey.

Generally, Denver's people were sympathetic. Blue was never charged with murder since the bodies upon which he and the others had fed were already dead. By December he was well enough to travel again. All dreams of quick riches had been forgotten. Daniel Blue returned to his native Whiteside County where he penned a pathetic account of his experiences on the Smoky Hill Trail. His narrative was first printed in 1860. But despite the stories of cannibalism, lack of water, and occasional Indian hostility, the Smoky Hill was easily the most popular route to Denver until the completion of the Kansas Pacific Railroad in 1870.

It was inevitable that with the discovery of gold, even more people would want to reach Pikes Peak. Since conditions west of the Missouri River left much to be desired, some easier way to get them to their destinations was needed. Because stagecoaches had long been used successfully in the East, it was only a matter of time before they would be introduced to the Great Plains. A few optimists were already pushing for railroads.

Soon there were three stagecoach lines serving the Pikes Peak towns. One came up from Santa Fe, while the other two entered from the East. On May 7, 1859, the first coach wheeled into Denver after six continuous days on the road. It belonged to the Leavenworth and Pikes Peak Express Company, which in turn was owned by the firm of Russell, Majors and Waddell of St. Joseph, Missouri. This pioneer enterprise had the jump on most of its competition, since it already operated freight wagons in the Pikes Peak region. It was easy for them to readjust their schedules and begin operating daily stagecoaches to Denver.

Russell, Majors and Waddell owned forty Concord coaches, hand-fashioned in far away New Hampshire. Later, their stable was expanded to fifty-two vehicles. They were generally painted red and cost eight hundred dollars each. Each one could carry nine tightly-packed passengers. For the crossing to Denver, each coach was drawn by four mules.

The first stage left Leavenworth at the head of a parade, serenaded by a brass band. It followed the Republican River road to its junction with the Smoky Hill Trail. On the return trip, the coach carried three thousand five hundred dollars in gold. Without incident, it arrived back in Leavenworth on May 21, 1859.

Later, the Leavenworth and Pikes Peak line was able to cut its rather remarkable six-day time to just four days and six hours for the crossing from Missouri to Denver. After just six weeks, Russell, Majors and Waddell gave up on the Smoky Hill road in favor of the longer but less risky Platte River road across Nebraska. This, of course, was the well-worn Oregon Trail, but in Nebraska, it has always been called the Great Platte River Road. In western Nebraska they cut south to Julesburg, then continued down the river to Denver.

At first the $125 ticket included meals. Later, as traffic increased, the charge went up to $200, one way, meals not included. Three times each day, the coaches stopped at a home station where food was served at rates ranging from 75¢ to $1.25 per person. There were no menu choices. The unappetizing fare consisted of dry bread or biscuits, rubbery bacon, half-cooked beans, and dishwatery coffee. Afterwards, one could use the community comb and tooth brush that were tied on strings outside by the mule trough, "to discourage theft." The welfare of passengers always took a back seat to care of livestock, since mules were expensive to replace. Personal baggage was retricted to twenty-five pounds for each passenger.

The coaches rolled day and night to keep up with the tight schedules. Many people ingested copious amounts of alcohol to soften the jolting ride and to bolster their courage. Nine passengers shared two uncomfortable bench seats inside the coach. Frequently, a broad-beamed person would be "persuaded" to ride topside with the driver. For those who had the price, this was the fastest and easiest way to Denver.

Thick horsehide straps, layered and secured with copper rivets, supported the coach body independently of the vehicle's frame. Lacking springs, this unique suspension system allowed the coach body to swing from side to side, finding its own pendulum-like center of gravity on sloping mountain roads. Although less comfortable than some of its eastern competitors, the Concord was also less likely to tip over, and consequently became the preferred vehicle on the less-refined western routes.

During warm, sunny days, the passengers sweltered; in rain storms, they were drenched; on dry days, the dust was suffocating. If the rolled leather curtains were let down, coach interiors became rolling saunas. There were no deodorants. Passengers found relief only during the brief stops at swing stations. There they could take deep breaths, share a bucket of water, and the same dirty roller towel, also firmly secured against pilfering.

Although much admired as the "best in the West," the drivers were frequently drunk and always profane. Working with mules, they insisted, had a peculiar but predictable way of expanding the human vocabulary. Most stagecoaches traveled in pairs because of the popular fear of trouble from Indians or the ever present possibility of accidents.

Russell, Majors and Waddell built a series of way stations across Kansas, Nebraska, and Colorado. They called them mile houses, according to how far out of Denver they had been built. There were twenty-seven between Leavenworth and trail's end at Denver. For example, the Four Mile House was built just four miles southeast of trail's end. It still stands at the

Golden Gate City, as it appeared at the time of the gold rush.

Here is the empty Golden Gate City site as it appeared in 1982.

end of South Forest Street, the restored center of a public park, refurbished and maintained by Denver's Parks and Recreation Department. Built originally of logs in 1858, the structure contained an eating facility, a bar, sleeping rooms, and an emigrant camp ground among stately cottonwood trees for the use of thrifty travelers. Four beautiful white horses were kept at the Four Mile House. These handsome animals were fondly recalled by migrants who had used this facility. In later years, wood shingles were facaded over the original square hewn log walls.

With thirteen rooms, the Twelve Mile House was the largest hostelry in the territory. Two rooms of the original Seventeen Mile House are still standing. It was hardly more than a bar within a log cabin. There was a Nine Mile House at one time. The two-story Twenty Mile House was at Parker. Of frame construction, it contained ten sleeping rooms. It also sported a large dining room on the ground floor, for the accommodation of hungry travelers. Other trails used the "Mile House" designation too. For instance, there was a Sixty Mile House at that distance west of Omaha on the Oregon Trail.

By 1860 the Leavenworth and Pikes Peak Express Company had been reorganized as the Central Overland, California and Pikes Peak Express Company, but it was still owned by Russell, Majors and Waddell. Another of this firm's enterprises was the romantic but ill-fated Pony Express experiment. Colorado's only Pony Express station was at Julesburg. A relay of riders used the Platte River Trail to bring letters from the East to Denver.

For a charge of five dollars riders would carry a half ounce letter through to Sacramento. Later this fee was reduced to two dollars, and only one mail was ever lost. Using modern cost accounting methods, we now know that the actual cost per letter was thirty-five dollars. At this rate Russell, Majors and Waddell were bankrupt in just sixteen months. They lost not only their beloved Pony Express, but also the stagecoaching end of the business as well. Ben Holladay, far-famed as the

"Stagecoach King," purchased what remained, incorporating their schedules into his Overland Mail and Express Company.

From Atchison, Kansas, to Denver, the stagecoach fare became seventy-five dollars, raised by Ben Holladay, who later doubled the price. In 1865 travel by stagecoach was resumed on the Smoky Hill road. This time, it was John Butterfield's Overland Despatch Company. But the effort was short lived. Butterfield also sold out to Ben Holladay. Reporter Bayard Taylor rode the last stagecoach to Denver on the Smoky Hill Trail in June 1866.

Here and there one may still see traces of the Smoky Hill Trail, mostly in the form of double ruts crossing a field. Present U.S. Highway 36 is very close to the original North Smoky Hill road. U.S. Highway 40 and 287 through Kit Carson and Hugo follow the South Smoky Hill route. Contemporary Interstate 70, east of Denver through Limon and Burlington, is the old Middle Smoky Hill or Starvation Trail.

# 8.

## SUPPLY TOWNS, THE JUMPING OFF PLACES

ALTHOUGH MANY mining rushes were dependent upon placer mining, in Colorado lode or vein discoveries were important after 1859. News of the rich Jackson and Gregory finds nearly depopulated the tiny Cherry Creek towns. Stores were boarded up, vegetable gardens were untended, while carpentry and construction work in progress was left unfinished as gold hungry optimists stampeded out of Denver.

Once the presence of gold in the mountains had been confirmed, a series of three small miner's supply towns grew up along the hogback at the eastern base of the foothills. They were located only a few miles apart and all three had the advantage of business houses that were closer to the gold fields than the Denver suppliers. From north to south, they were named Golden Gate City, Apex, and Mount Vernon.

Golden Gate City was situated at a point slightly northwest of present Golden, beside the junction of Colorado Highway 93 and the Golden Gate Canyon road. This canyon took its name from one of two circumstances. First, it was supposed to be the gateway to the golden riches that lay beyond it in the Central City area. The second version is that it was named because it was located at the same latitude as the famed Golden Gate at San Francisco, California.

When Golden Gate City was first established by the Golden Gate Town Company in July 1859, town lots were free. It consisted mostly of tents. These were soon replaced by a double row of about fifty log structures. In the round numbers so beloved by 19th century statisticians, a thousand men and seventy women made up the initial population. More than half

From North Table Mountain in spring, the arrow indicates the scar of Golden Gate City's main street.

In this view, Evelyn Brown leads hikers on the toll road grade above Golden Gate City.

Library, State Historical Society of Colorado

This early view of the residents of Mt. Vernon shows little of the town.

From the Collection of Evelyn and Robert L. Brown

By 1982, most of the Mt. Vernon location was under the landfill of a freeway.

of them were somehow involved with the business of selling a wide assortment of supplies to transient miners. A toll gate stood at the north edge of town. In early spring, when the afternoon sun is low in the sky, the original scar of Golden Gate City's single street can be seen if you look down into the valley from North Table Mountain.

The second supply town of this cluster was known as Apex. It was located directly east of the mouth of the steep canyon behind the Heritage Square amusement park. More precisely, its site is now buried under the landfill that supports the miniature railroad concession at the north end of the park. This earlier town of Apex should not be confused with the 1890s mining camp of the same name in Gilpin County.

Less than a mile south of Apex stood the town of Mount Vernon. Nearly all of its site is now buried by the landfill that supports Interstate 70. The precise location was barely southwest of the point where the interstate crosses the Morrison Hogback Road, which is County Highway 93. The Matthews Winters Historic Park currently occupies a tiny portion of the old townsite. Robert W. Steele, first governor of the illegal Jefferson Territory, lived at Mount Vernon. His cabin is now identified by a granite marker. Steele also had a second cabin at nearby Apex. He moved back and forth as his moods dictated. The single surviving stone house was built and owned by George Morrison. Because he was in the quarrying business, Morrison naturally built his home of stone. In 1860 the first schools in the territory were in session at Denver, Golden, Boulder, and Mount Vernon. A town plat for this once hopeful community still exists in the Jefferson County courthouse.

From each of the three supply towns, access roads of a sort led up to the better mining regions. Since Colorado had no highway department in 1859, all of these routes were toll roads, paid for by those who used them. From Golden City, the Golden Gate and Gregory Toll Road was incorporated in 1862. It extended above the twisting, serpentine canyon to the top of Guy Hill. The original gold seekers route was well above the contemporary grade and followed the ridge tops.

From Guy Hill, it ran downgrade into Black Hawk. Today a county road follows the curvy lower contours of Golden Gate Canyon.

Supposedly, there are still scarred trees along this section caused by rope burns. Early wagons had inadequate braking systems. On steep hills, one end of a stout rope was secured around the wagon's rear axle. The opposite end would then be wrapped around the trunk of the largest tree at the top of the hill. Two or three turns were usually adequate. With this system, one man could let the rope out bit-by-bit, lowering the wagon safely down the incline.

Beyond Apex, the steep grade of Apex and Gregory Toll Road climbed straight west up the canyon. It continued on through the present Paradise Hills subdivision, turning north at a point just east of the Mount Vernon Country Club complex. Early newspaper accounts advised people to avoid the Apex road because the odor of dead animals was so unpleasant. At a later time, the road intersected with the Chimney Gulch Toll Road and went north. Still another branch turned south down the gulch from present Paradise Hills to the Denver City, Mount Vernon and Gregory Toll Road.

Originally, the Chimney Gulch Road was a wagon route, built and operated as a toll road between Golden and Lookout Mountain. The toll gate and keeper's house were located at the bottom of the gulch. When fire destroyed these landmarks, only the stone chimney of the house was left standing, hence the name Chimney Gulch. At the Forks of Clear Creek, one had the choice of going north to the so called "Gregory Mines" or south to the Jackson Diggings.

From the town of Mount Vernon, the third of the principal roads, The Denver City, Mount Vernon and Gregory Toll Road entered the mountains. It too followed the spine of a gulch west of the town. Near the summit of the hill, beyond Ruby Station, the Mount Vernon road cut off to the north at Hayward's Junction, intersecting with the steep trail down Big Hill into Clear Creek Canyon. Every road had a difficult hill. E. W. Henderson of Central City was the first to negotiate this road

*From the Collection of Evelyn and Robert L. Brown*

The foreground scar is the old Chimney Gulch Toll Road. Table Mountain and Golden are in the middle distance.

*Collection of Freda and Francis Rizzari*

Long after the Gold Rush, this final stage was photographed on the Chimney Gulch road in 1917.

by wagon. He snubbed it down the precipitous grades with a block and tackle. The road was so bad that twenty yoke of oxen were required to transport a small boiler over the road, blocking the trail for other traffic.

Just above the Clear Creek Canyon road, west of tunnel Number Three but east of the Forks of Clear Creek, there are more examples of trees scarred by braking ropes. To find them, cross over to the south side of the creek (carefully), and follow the dim trail up the hill, watching for the largest and oldest trees. The scars are at eye level height. Keep in mind that trees grow from the trunk out and from the top up. Therefore, the abrasions are still found at their original height.

If you wish to view this phenomenon, wait until early October when the water level is lowest in Clear Creek. Although the scars of the Big Hill Wagon Road are easily seen from the highway, getting to the other side of the creek is neither a pleasant nor a safe undertaking. The rocks are slippery and the current is swift. As the trail enters the upper reaches, it becomes nearly impossible to find its traces in the Mount Vernon Country Club area.

Inevitably, other new roads had to be constructed to connect the newer gold towns of South Park with the older Cherry Creek settlements. West of Denver, the route to South Park followed an old Indian trail over 9,950 foot high Kenosha Pass. The original Kenosha Pass crossed the ridge at a point about a half mile east of the present road and made a much steeper, more direct descent into South Park.

A second road crossed Guanella Pass from South Clear Creek to the North Fork of the South Platte River. Then the two roads combined, crossing over Kenosha Pass to the placer gold towns of Hamilton and Tarryall. Altogether, this road was much too rough, and at times was nearly impassable. By 1860 Kenosha Pass had been improved. Later, both the Barlow and Sanderson and the McLaughlin Brothers stagecoach lines used this low crossing from the South Park towns to Denver.

Another early road went south along the Front Range to El Paso. From there, it followed the original Ute Pass, which

## MOUNT VERNON

is situated on the Denver, Mt. Vernon and Gregory Toll road, at the entrance of the canon, where the road enters the mountains, fifteen miles from Denver. It has a population of about 200. Lime, and the crude material for making Plaster of Paris, is found in great abundance; the finest of stone for building purposes, having the appearance of common marble, is obtained here in inexhaustible quantities.

# DENVER CITY, MT. VERNON

— AND —

# GREGORY

TOLL ROAD.

THIS IS THE

## SHORTEST, BEST AND MOST TRAVELED ROAD

### TO ALL PARTS OF THE MINES.

The Roads are always kept in good repair, and

## TOLL CHEAPER THAN BY ANY OTHER ROAD.

### OFFICERS OF ROAD.

GOV. STEELE, Pres.                    J. C. NELSON, Vice-Pres.

JAMES GALBRETH, Secretary.

# MOUNT VERNON HOUSE,

## Mount Vernon, Rocky Mountains.

# G. MORRISON, - - - PROPRIETOR.

This House is now ready for the reception of guests, and the patronage of the Traveling public is respectfully solicited.

*Collection of Fred Rosenstock*

This ad extolled the virtues of the Denver City, Mt. Vernon and Gregory Toll Road.

Gold Hill was an important town in early Boulder County.

Gold Hill still contains many of its original structures. Inset shows token from Frank V. Boyd general store.

was then little more than an Indian migratory path. It traced the spine of the first valley south of U.S. Highway 24, the present Ute Pass Road. The old original trail still exists and can be hiked for several hilly miles out of Manitou Springs. Double-rutted wagon tracks are visible at several points along its right of way. The trail begins behind the parking lot of the Mount Manitou Incline, climbs steeply uphill, then plunges into a deep valley. Up and down, the tiresome pattern is repeated several times in the first few miles. Little wonder that the more gradual U.S. 24 was constructed as a replacement, Augusta and H. A. W. Tabor, among others, followed this older Ute Pass to reach South Park and the upper Arkansas Valley.

Not a few of those who chose the steep grades of Ute Pass were forced to pack their supplies on the backs of mules. The trail was transformed into a gold rush wagon road when residents of the early town of El Paso reponded to rumors of gold in South Park. Beginning in January 1859, they loaded ox-drawn wagons and started up the hilly route. By shoring up the narrow places and with much road bulding, they actually reached Fairplay and were rewarded with a good placer gold find in the South Platte River. However, a hard winter dictated abandonment.

Ute Pass became a free road, widely touted by El Paso businessmen as the shortest, finest, and most level road to South Park, as well as to the gold placers at Oro City on the upper reaches of the Arkansas River, and to the Blue River discoveries on the western side of the Continental Divide.

But little El Paso never realized its dream of becoming the prime supply town for the upland settlements to the West. In fact, it was nearly abandoned before it re-emerged as Colorado City in 1859. The precise El Paso location would be the western part of Colorado Springs and the eastern edge of Manitou Springs.

# 9.

## OVER THE MOUNTAIN WALL

AT FIRST the gold hunters had hovered around the initial mining districts of Clear Creek and Gilpin Counties, and in Boulder Valley. Gradually, these areas began to fill up; opportunities for making fresh new finds were diminished by sheer numbers of people. Since the mountainous areas were now recognized to be the real sources of gold, it was inevitable that the rush must spill over into them.

When warm weather returned in the summer of 1859, long lines of wagons could be seen moving out of Denver and the other Piedmont towns. All lines led west in the direction of the growing profusion of mountain mining camps. For the less adventuresome, there were a number of small satellite communities growing up around Central City. Not far away, west of Boulder City, there were some new mining camps around Gold Hill. One of the first mining districts to develop was in Boulder County.

In January 1859, a party of gold seekers from Nebraska camped at the Red Rocks area near the mouth of Boulder Canyon. Where they raised color in Boulder Creek, a place called Gold Run soon enjoyed a flurry of popularity. On the hill above, the source was found to be a rich gold vein, and a town called Gold Hill was started. In June of that same year, another strain of rich gold was located in the Horsfal lode, also close to Gold hill. A modest rush into the Boulder district followed.

The Nebraska group consisted of twelve men under the leadership of Capt. Thomas Aikens. Soon eleven log cabins had been erected for winter shelter. The Nebraskans assured

Collection of Mrs. Dean Ives

In this early view of Montgomery, the stage road to Hoosier Pass can be seen in the background.

From the Collection of Evelyn and Robert L. Brown

From the same angle, Montgomery's site now lies beneath this reservoir.

English-speaking Chief Left Hand that they would leave for the mountains to prospect in the spring.

Another 1859 Boulder County discovery was made by B. F. Langley who found what he called Deadwood Diggings in Gambell Gulch, a tributary of South Boulder Creek. Contrary to some published reports, the name came not from the famed Black Hills gold camp, which didn't even exist until 1879, but from the vast amount of fallen timber that littered the spine of the gulch.

Many among the incoming multitude of adventurers looked across the Front Range toward South Park. When discoveries were made there, a part of the rush spilled over into that high upland region. As news of the South Park discoveries reached Denver in July, gold-greedy optimists flooded westward into these newer districts. Although a few came in 1858, most arrived in 1859.

Once again a collection of camps and towns was thrown together at locations close to the richer finds. Near the eastern foot of 11,482-foot Breckenridge Pass stood the placer mining towns of Silverheels, Hamilton, and Tarryall. The latter two towns were situated in Tarryall Gulch, but on opposite sides of Tarryall Creek, some six miles east of the Continental Divide. Breckenridge Pass became Boreas Pass in 1882 when the railroad was built across it. The Tarryall diggings were nicknamed "Graball" by late comers who found no space left for them.

Roughly five thousand people had surged into that district alone as early as the summer of 1860. Since these were essentially placering camps, the creek banks were literally lined with whipsawed sluice boxes. Several good pockets of shot were also opened up that first summer.

Hamilton was a long, narrow town, built north of the fork where the road over Breckenridge Pass split off from the main route that continued across South Park to Fairplay. Hamilton had only one principal street, which extended back into a deep pine forest. There were no sidewalks. In 1860 the Rev. William Howbert began, but did not complete, a Methodist Church building for the town.

*Collection of Freda and Francis Rizzari*

In South Park, Alma, Colorado, looked like this in gold rush days. Mount Bross
is the background peak.

*From the Collection of Evelyn and Robert L. Brown*

In modern Alma, house at left is the same as in the above photograph. Inset shows
Alma tokens.

Many saloons stood along the principal thoroughfare, but there were few stores. Sometimes a man would ride directly up to the front of a store and give his order without ever leaving the saddle. Not many stores were needed at first, as most people had arrived in the Rockies with a six months' supply of food in their wagons. Few of them had seriously considered staying longer.

At the western extremity of the park, nestled in the folds of the Mosquito Range, were the communities of Dudley, Alma, Fairplay, Quartzville, Buckskin Joe and Montgomery.

Across the Mosquito Range from South Park, placer gold had been found in California Gulch near the headwaters of the Arkansas River. A town called Oro City had already appeared there by 1860. French and Georgia Gulches attracted others into present Summit County. Towns like Lincoln City and Parkville mushroomed in the high valleys near the gold placers of Breckenridge. Others appeared almost overnight in the nearby Blue River Valley.

In August 1859, more gold was uncovered in the Swan River Valley, causing yet another surge of humanity into present Summit County. Both placer and lode gold were found in the upper tributaries of the river. By far the richest of the placer gravels was found at depths of from four to twenty feet, close to bed rock.

Since the Loveland Pass road existed only as an Indian path at that time, miners cut a road across the Continental Divide from the Georgetown end of South Clear Creek. It followed Leavenworth Creek southwest, crossing the above timberline slopes on a precarious shelf at 13,132-foot-high Argentine Pass. On its opposite side, the road dropped precipitously down into the valley of Peru Creek, then on to the valley of the Snake River. First called Sanderson Pass, when it was completed in the 1860s, its name was soon changed to Snake River Pass, then finally becoming Argentine Pass. Although this historic road is impassable for all but foot travel on its western side, it remains the highest Continental Divide crossing on the North American continent. More importantly, it provided ac-

*Collection of Nancy and Ed Bathke*

In South Park, Fairplay was the most prominent town.

*From the Collection of Evelyn and Robert L. Brown*

Contemporary Fairplay has a resident population of about 500 people. Inset shows bar tokens from Fairplay.

*From the Collection of Evelyn and Robert L. Brown*

Here was Breckenridge, the early gold town in Summit County. Note Denver Hotel.

*From the Collection of Evelyn and Robert L. Brown*

Here is contemporary Breckenridge. Inset shows Denver Hotel Bar token and one from Ed Keller's saloon.

cess to the Swan River Valley, Breckenridge, and to the Summit County mines generally.

Although the Civil War period brought about a severe slump in Colorado mining, by the time it ended in 1865, the Pikes Peak country had become a far different place than it had been in the earlier gold rush days. The easy surface gold was gone, and people did not yet know how to extract the deeper deposits. By the time the veterans of our fraternal struggle had trickled back to Colorado, Denver had become the most important community. In the end, it prevailed over its numerous rivals, absorbing many of them and becoming the most enduring of the Piedmont settlements.

# 10.

## SOCIAL AND CULTURAL ASPECTS OF THE GOLD RUSH

INEVITABLY, the migratory tides washed in various proportions of unskilled laborers, drifters, boomers, ne'er-do-wells, drunks, gamblers, prostitutes, and an infinite variety of other human flotsam. Many of the others were incredibly fine people who were down on their luck. Those men who came were either bachelors or married men who had left their wives and families in the East until they could get established in the new territory.

For most of them, getting a job was easy, but finding adequate, affordable housing was nearly impossible. The shortage of women condemned most of the men to single blessedness by simple arithmetic. Normal male-female relationships were thrown completely out of balance. Some temporarily separated husbands went in search of other female companionship. Some others threw caution to the winds and plunged more deeply into matrimonial waters by marrying again, if they could find a partner. Needless to say, prostitution flourished in nearly all of the towns.

As the new towns grew, so did their problems. Soon miners found it expedient to engage full time law enforcement officers. It was generally conceded that anyone with good feet could become a policeman. Mainly, their duties consisted of running off unlicensed peddlers seeking to compete with local merchants, forcing Chinese laundrymen to close on Sundays, and hauling owners of wandering pig herds before a Justice of the Peace.

Law and order in a mining camp society was provided by miners' courts and claims clubs. Members often supported the authenticity of each other's claims with ropes and firearms.

Since most of the early settlers had come primarily for the gold, there were very few plans for a permanent society. The idea was to get rich quick, return home, and spend whatever was left. Consequently, there was a general reluctance to put out money for civic improvements, jails for instance. So, when a miners' court found someone guilty of a crime, the penalties assessed were both economical and severe.

For minor infractions such as perjury, petty theft, or public drunkness, Denver had a whipping post. Most people owned whips, so the investment was slight. Ten to twenty lashes were the common sentences. For more serious violations of good taste, like wife beating, assault, major theft, or similar offenses, the penalty was banishment from the district. Walking the miscreant to the town limits at gunpoint was not a costly process. For murder or theft of a mule or horse, the punishment was the inexpensive, ultimate pronouncement — hanging in a public place. There were no appeals, pardons, and no probation in miners' courts. If found guilty, defendants were "taken to the tree." One jurist recalled that they never hanged on circumstantial evidence, but if the accused was found guilty, all jury members held the rope and pulled together when the signal was given. In most instances, the source of irritation was immediately "jerked to Jesus."

Denver's first official hangman was Thomas Pollock, who doubled as the town's blacksmith. At other times, the eclectic Pollock served as Denver's marshal and occasionally as the undertaker. Locally, he was known as "Noisy Tom," for his habit of running through the streets prior to one of his executions, exhorting one and all to "come and see the hanging."

Another marvelous character who appeared at Cherry Creek was Charles Gardner, a huge, unkempt man known locally as "Big Phil the Cannibal." In the 1840s he had escaped from a Pennsylvania prison where he had been incarcerated for murdering a Catholic priest. At the time of the gold rush he was living with a band of Arapaho. At other times, he could be found bumming drinks in early Denver's saloons. His pecular sobriquet came from the fact that he was from Philadelphia and the circumstance of his having eaten one of his Indian wives during a particularly hard winter.

Gardner was also credited with having devoured a male Indian, a white man, and a Frenchman. Once in the aftermath of an Indian attack, Phil was thought to be dead and was scalped. The hairless circle on the top of his head never quite healed. One of his great pleasures was to bow deeply while removing his hat as proper ladies approached. Most members of the fair sex could be counted on for a good scream, while others obliged by fainting. Either way, Phil was delighted.

In early Denver, a criminal gang called the Bummers attempted to dominate the townspeople, exploiting them for their own nefarious purposes. They were led by Charlie Harrison, whose Criterion saloon became their headquarters. When the citizens finally got their act together, an opposing group — vigilante in nature — was formed. They were known as the Stranglers. Fortunately for Denver, the latter group prevailed and things settled down.

But the effects of Colorado's gold rush on its emerging society were by no means resticted to the growth of law enforcement nor of a precious metals mining industry. It has been estimated that for every working miner, five service people were required to provide such necessities as food, entertainment, government, general merchandise, liquor, newspapers, transportation, and other civilizing elements. About the only items that the working miner could provide for himself were firewood, venison, and a very limited number of hardy vegetables.

Most of the service people were fully as transient as the miners they served. However, some members of both groups remained after the initial excitment had subsided, thus providing the nucleus of a permanent society. But it had not been planned this way. A majority of those who had succumbed to the lure of easy riches planned to return home after they had made their pile, giving the country back to the Indians. But when the time came, many were reluctant to leave; others lacked the price of an outfit or a ticket on the stage.

When the first census of Arapahoe County, Kansas, was

taken in 1860, some 34,277 persons were still there. An additional four thousand people who had settled around Boulder were included in Nebraska's count. Those who had come west only to return home in 1859, the "go backs," could not be counted.

One of the surest signs that civilization and culture had arrived on the frontier was the appearance of a church. No more positive indication of a town's permanence existed. To symbol-conscious easterners, the presence of a house of worship meant that here was a community that cared, a town that was family-oriented to which one might bring wives and children. Bells, although costly, might be hung in the steeple to enhance the effect. Because they were homesick for familiar eastern religious institutions, most early frontier churches were painted white. Glass windows were scarce, stained glass was unobtainable.

In the beginning, eastern missionary societies had feared that the West's "Godlessness" would spread and that irreligion would sweep back across the Missiouri River. So, they recruited missionaries, called circuit riders, to save souls as they traveled throughout the West, preaching wherever a group of men and women could be found to listen. These dedicated persons included two Methodists, the Reverend W. H. Goode and the Reverend Jacob Adriance, who were the very first men of the cloth to arrive in the gold fields. The Reverend William Howbert, another Methodist, brought his family west in a wagon during the initial surge of gold rushers. The Reverend Joseph Machebeuf was sent north into Colorado by the Bishop of Santa Fe to minister to those of the Catholic faith.

A third Methodist, orginally a lay preacher and carpenter, the Reverend George W. Fisher was the first to conduct services in the gold camps. Fisher was a member of Larimer's original town company. He held the first service in one side of the Russell duplex cabin in Auraria. Twelve miners and two curious Indian women made up his initial congregation. Although services were proceeding in Russell's end of the cabin, gambling sounds were clearly audible from the other side of the wall, occupied at that time by squaw man John Smith.

Another account of this first service, long popular in Denver, insists that Fisher actually preached in Smith's saloon, beginning his sermon with, "Everyone who thirsteth, come ye to the waters. Come buy wine and milk without money." Behind his improvised pulpit hung an ironic sign that advised, "No Credit — Pay as you Go."

In January 1860, the Reverend H. J. Kehler, an Episcopalian, arrived with three daughters, a granddaughter, and a seventeen-year-old son. A popular folklore account tells us that the Reverend Kehler was given the use of a room above Ed Jump's saloon. There, due to the gambling and drinking sounds that penetrated to the upstairs, hearing the sermon was all but impossible. Finally, Jump agreed to suspend gambling for one hour, a sacrifice he firmly believed would gain him entrance into heaven. Actually, the Reverend Mr. Kehler held the service in Jesus Abrieu's cabin that stood in the bed of Cherry Creek. At other times, the Abrieu abode was used as a community reading room.

Perhaps the lot of the circuit rider can best be appreciated by recounting some of the story of the Methodist Reverend John L. Dyer. In Colorado, he was widely known as "Father" Dyer, because of his age. When assigned to come here, he walked for seven hundred of the miles from Minnesota, reaching Denver on June 20, 1861. He was sent first to minister to the miners at Buckskin Joe in South Park. Dyer often walked the seventy-six miles from Fairplay to Denver in order to save the $2.45 coach fare. By the end of 1861, he had walked over five hundred miles, receiving $43 in collections. His salary from the church was next to nothing. To earn extra money, he contracted to carry the mail from Buckskin Joe across the thirty-seven miles of Mosquito Pass to Cache Creek. On Sundays he preached. He was paid $18 weekly for one round trip. In winter he used Norwegian snowshoes (skis), acquiring the nickname "The Snowshoe Itinerant."

Following a stint of preaching on the plains, Father Dyer was sent back to the South Park district in 1876. Three years later, his circuit included Dudley, Carbonateville, Brecken-

The Reverend John L. "Father" Dyer was one of the most effective of the early circuit riders.

ridge, Kokomo, Robinson, Decatur, Fairplay, Buckskin Joe, Alma, Montgomery, and Mount Lincoln. He retired from active preaching in 1880. Dyer died at his Denver home in 1901 and was buried beside his family in Castle Rock.

To encourage clergymen to stay in town, Denver donated four building lots each to the Presbyterians and Methodists, ten lots to the original Hebrew congregation, eight to the Catholics, sixteen to the Episcopal group, and six to the Baptists.

With some regularity, other signs of the arrival of civilization and permanence began to appear throughout the Cherry Creek towns. A marital mix-up in Denver resulted in the first divorce, with the family's quitclaim deed to real estate being awarded to the husband.

Another sure sign of permanence and progress was the arrival of more sophisticated equipment for the separation of gold from rock. Called stamp mills, in 1859 there was only one in the territory, at Nevadaville. By the end of 1860, another one hundred and fifty ore crushers had been packed in across the Kansas prairies on ox-drawn freight wagons. Their arrival signaled the end for pick-and-pan miners. It took eastern capital to mine deeply into the earth and supply the volume of ore needed to run the hungry mills.

Below town, the Reynolds course was established for horse racing. In the Apollo Hall, a saloon and variety theatre, a stage was constructed upstairs over the barroom. Upon it a traveling troupe of thespians blundered through an occasional siege of Shakespeare's *Richard III*. For more popular tastes, the players offered *The Flying Scud, East Lynn,* and other melodramas. Candles illuminated the stage while an appreciative audience of a few hundred sat uneasily on rough-hewn pine benches.

Denver's first cemetery, present Cheesman Park, was started by William Larimer, who sold it to the local undertaker, a thrifty soul who preferred to conduct his interments when nobody was watching. This enabled him to dump the bodies, thus saving the caskets for the next customer. He once placed a sign on his door that read, "Gone to Bury My Wife — Back in a Half Hour." He also separated burial plots of Roman Catholics from Jewish. Protestants had an area of their own too. Criminals were buried in a "Boot Hill," a common practice for the time.

Gold dust was the principal medium of exchange. Most people carried a buckskin pouch of dust gold. Merchants kept sets of small scales adjacent to the cash drawer. A pinch of gold was valued at twenty-five cents; it was considered good form to allow the clerk to take his own pinch from your pouch. Men with large thumbs were much sought-after by merchants in general and saloon keepers in particular!

In July 1860, Clark, Gruber and Company arrived in Denver City from Leavenworth, Kansas. They established a branch bank at sixteenth and Market Streets. Since private mints were legal at that time, they began casting ten dollar gold pieces. Later they fabricated $2.50, $5.00, and $20.00 pieces.

Despite the rowdy, hurly-burly life style of the miners, there was also a strong disposition toward Puritanism in the later mining camps. Some towns like Gold Hill and Saints John forbade saloons and gambling halls in their districts. Other areas refused to let attorneys practice in miners' courts.

From the very beginning, food was a problem for the easterners who came to Colorado. In the mountainous mining regions, cold nights and short growing seasons made agriculture difficult. On the plains, the practice of agriculture was frowned upon by the Indians, who insisted that it was women's work. Chief Left Hand said that farming was "a sedentary and unnatural occupation." Therefore, fresh vegetables could not be purchased from that source.

Charlie Autobees was a halfbreed French trapper from St. Louis. He had been an active mountain man during the earlier fur trade period. Upon its decline, he was one of those who chose to remain in the West, supporting himself as a farmer. Autobees was already plowing and planting in the Huerfano Valley prior to the gold rush. He sold corn to those early settlers who had entered the territory by way of the nearby Santa Fe Trail. Autobees was also the leader of a group that started a Mexican settlement near the mouth of the Huerfano River. It had become a fairly well established village by the time of the gold rush.

Well within the first year, substantial numbers of farmers had joined the rush westward. Some of them stopped at Highlands, planting truck gardens along Federal Heights on Denver's northwest side. Others took up homesteads in the lower valley of the Arkansas River and along several of its tributary branches. Corn, beans, squash, and other vegetables were soon transported to mountain towns for sale to the prospectors. Crops of wheat, rye, and barley followed. John W. Smith erected grist mills in both Denver and Boulder. All along the Platte Valley, on both sides of Denver, farms amd ranches were being established. Cattle, hogs, turkeys, and chickens were raised in abundance.

Very early in 1859, David K. Wall planted Golden's first vegetable garden on two acres. He realized over $2,000 for his crop. Seven acres were planted in 1860, producing $8,000 income from his miner customers.

On the future site of Denver's Overland Park, an agriculturist named Potato Clark raised abundant harvests of that vegetable. Clark was married several times. One of his brides was very proud of her long red hair. When the Platte River flood swept down the valley, her hair somehow became entangled in a barbed wire fence and she was drowned.

Late in the spring of 1859, Denverites and Aurarians were treated to a spectacular race between two competing newspapers to determine which would be the first to publish. Since so few people lived there, it was felt that only one paper could survive. It was understood by the contestants that the loser would drop out of the picture, leaving the field to the victor.

John L. Merrick of Independence, Missouri, arrived on the scene first. He brought a printing press that had once belonged to a party of Mormons. He planned to publish a sheet called the *Cherry Creek Pioneer*. William Newton Byers of Omaha, Nebraska, entered town on April 20 with his two outside pages of type already locked in their chases. With John Daily and Thomas Gibson, he rented the upstairs loft of Uncle Dick

*Western History Department, Denver Public Library*
William Newton Byers, as he appeared in his later years.

Wooten's Western Saloon at 1411-15 Eleventh Street. There, they hasily assembled their Washington press. Byers' paper would be called the *Rocky Mountain News*.

As word of the rivalry spread, the race began. Bets were made and raised. Each paper had adherents in its cheering section. Judges, armed with gold watches, tried to appear profound as they shuttled back and forth between the opposing establishments. The date was April 22, 1859. A light spring snow was falling outside on the anxious partisans. To complicate matters on the inside, Wooton's shingled roof was leaking on the *News* printers. Just after 10:00 P.M., John Daily removed a single sheet from the press, folded it and handed it to Editor Byers. The race was finished; the *News* had won by a mere twenty minutes. In the beginning, Byers' paper was published weekly. Subscribers were charged twenty-five dollars a year. Single copies sold for twenty-five cents.

Merrick's *Pioneer* was published, too, but its first issue was also the last. His press was sold to a southern sympathizer in Gilpin County. During the Civil War, it was used to print a pro-Confederate sheet in Central City. As for John Merrick, the money he received for his press was used to purchase a mining outfit. Merrick became a prospector. From time to time, as his resources became exhausted, he returned to the city and worked for Byers to earn new grubstakes for himself.

Schools, like churches, were regarded as bellwethers of a growing civilization. Denver, Mount Vernon, Golden, and Boulder all had institutions of learning in operation before the end of 1860. The person who opened the first school was Professor Owen J. Goldrick, who entered the Cherry Creek settlements in formal attire, complete with spats, a stovepipe hat, yellow kid gloves, and only fifty cents in his pocket. Local tradition persists that Goldrick's rather theatrical entrance was enhanced by his ability to swear at his oxen in Latin, thereby demonstrating that he was a learned person. Considering the high rate of mining camp illiteracy, one wonders who in pioneer Denver would have comprehended Latin, profane or otherwise. Although mostly illiterate, they were no doubt impressed by

Goldrick's entry. Despite marginal poverty, Goldrick soon arranged for the use of a single story log cabin in which he started the first establishment of learning. He called it the Union Day School. If one wished to enroll a youngster, the tuition was three dollars monthly.

A source of considerable concern and no little irritation was the erratic delivery of mail from the East. Whether you could read or not, getting a letter from home was a major event. Fort Laramie, some two hundred miles north on the Oregon Trail, was the closest of the early postal facilities. Once each month, an "express" left letters and papers there. The following day, Jim Sanders and his Indian wife would drive their buckboard north from Denver to Fort Laramie, returning with an average of seventy-five to one hundred pieces of mail. Sanders charged twenty-five cents for each item that he relayed to Denver.

On February 11, 1860, Denver City acquired its own post office. As noted previously, for brief intervals the mail had been received at the offices at tiny Coraville and at Auraria. But with the coming of its own postal facility, a sense of true permanence came to the Cherry Creek town.

Inevitably, politics, both territorial and national, came to influence life on the frontier. The Presidential election of 1860 was a strange contest. First of all, the Democrats rejected their own sitting chief executive, James Buchanan, and steadfastly refused to renominate him. Northern Democrats, called "Regulars," nominated Stephen A. Douglas, while the Southerners, called "Bolters," rallied behind John C. Breckinridge of Kentucky. With two Democrat tickets in the field, the party's vote was split along North-South lines. Republican candidate Abraham Lincoln, with a mere forty percent of the votes, coasted into office over the divided opposition.

On February 29, 1861, Congress voted statehood for Kansas, but decreased its size by establishing its present western boundary, thus deleting its western counties that are now the eastern part of Colorado. Retiring "lame duck" president Buchanan signed the bill creating the new Colorado Territory on February

28, 1861. Four days later, he was out of office. It remained for Mr. Lincoln to appoint the first territorial officials. For governor, he chose William Gilpin, a Lincoln friend and long time Colorado enthusiast. Gilpin had been practicing law in Independence, Missouri. He was only forty-one years of age when the president appointed him to the Colorado post.

Between 1861 and 1865, the Civil War sharply divided as well as depleted Colorado's population when loyal sympathizers returned to their respective homes to defend their regional viewpoints. And, our fraternal struggle cost Governor Gilpin his job. When faced with a full-scale Confederate invasion from Texas, intent on capturing Colorado's gold, Gilpin, pleaded with Washington for military assistance. President Lincoln told him that no troops could be spared. So the governor issued drafts on the U.S. Treasury to defray the costs of his militia, the first Regiment of Colorado Volunteers. At Glorietta Pass, New Mexico, the Coloradoans decisively defeated the Confederates in a series of three spectacular battles on March 26 and 28. Despite the wisdom of Gilpin's act, the drafts were illegal. Those persons who cashed them demanded and got the governor's removal.

Lincoln replaced Gilpin with another personal friend, a man equal to his predecessor in every way. Dr. John Evans was the founder of Evanston, Illinois, of Northwestern University, and of its medical school. He started Chicago's series of public parks and its public school system. In Colorado, he built railroads and founded the first institution of higher learning, Denver Seminary, today's University of Denver. Unfortunately, the Evans administration faced a period of Indian militancy as the plains tribes took advantage of involvemnt with the Civil War.

Like Gilpin, Evans pleaded his case in Washington, but recieved no troops. While he was gone, Col. John M. Chivington led the Third Regiment of Colorado Volunteers to Sand Creek, where he attacked a Cheyenne village. The Sand Creek Massacre (or battle) is one of the most controversial events in Colorado history. Since Evans was unable to "solve" the Indian problem, he too was removed from office.

*Library, State Historical Society of Colorado*

William Gilpin, long time western enthusiast, became the first governor of Colorado Territory.

John Evans replaced Gilpin in April of 1862, becoming Colorado's second territorial governor.

When the war was finished a new migration, mostly family-oriented, surged into the new territory. These latest arrivals looked at their new mountain environment optimistically, sure that Colorado would bestow material and personal rewards upon them. West of Denver, a profusion of new settlements of varying sizes grew up adjacent to the several mining districts that were sprinkled within and beyond the Front Range. Existing stagecoach routes were expanded and changed to accommodate the growing network of cities and towns.

By 1865 schools and churches were no longer a rarity and could be found in one form or another in most of the camps. Likewise, the beginnings of stable governments were evident nearly everywhere. The gestation period was drawing to a close.

With a measure of stability came permanence. Many of the people who had stayed through the war years looked around them, liked what they saw, and decided to cast their lots with this new country. As a later consequence of the Pikes Peak Gold Rush, the new state of Colorado was created in 1876. The evolutionary cycle had been completed, the foundation was firmed up, and the Pikes Peak region would never be a frontier again.

# INDEX